U. Schnars · W. Jueptner

# Digital Holography

Ulf Schnars · Werner Jueptner

# Digital Holography

Digital Hologram Recording, Numerical Reconstruction, and Related Techniques

With 100 figures

Springer

**Dr. Ulf Schnars**
Im Grund 7
D-27628 Hagen
*schnars@t-online.de*

**Prof. Dr. Werner Jueptner**
Bremer Institut
für angewandte Strahltechnik (BIAS)
Klagenfurter Str. 2
D-28359 Bremen
*jueptner@bias.de*

ISBN 3-540-21934-x Springer Berlin Heidelberg New York

Library of Congress Control Number: 2004111072



Springer. Part of Springer Science+Business Media
springeronline.com



Typesetting: data delivered by authors
Cover design: medionet AG, Berlin
Printed on acid free paper 62/3020/M - 5 4 3 2 1 0

# Preface

Sag' ich zum Augenblicke
verweile doch, Du bist so schön
J.W. v. Goethe, „Faust"

An old dream of mankind and a sign of culture is the conservation of moments by taking an image of the world around. Pictures accompany the development of mankind. However, a picture is the two-dimensional projection of the three-dimensional world. The perspective – recognized in Europe in the Middle Ages – was a first approach to overcome the difficulties of imaging close to reality. It took up to the twentieth century to develop a real three-dimensional imaging: Gabor invented in 1948 holography. Yet still one thing was missing: the phase of the object wave could be reconstructed optically but not be measured directly. The last huge step to the complete access of the object wave was Digital Holography. By Digital Holography the intensity and the phase of electromagnetical wave fields can be measured, stored, transmitted, applied to simulations and manipulated in the computer: An exciting new tool for the handling of light.

We started our work in the field of Digital Holography in 1990. Our motivation mainly came from Holographic Interferometry, a method used with success for precise measurement of deformation and shape of opaque bodies or refractive index variations within transparent media. A major drawback of classical HI using photographic plates was the costly process of film development. Even thermoplastic films used as recording medium did not solve the hologram development problem successfully. On the other hand the Electronic Speckle Pattern Interferometry (ESPI) and it derivate digital shearography reached a degree mature for applications in industry. Yet, with these speckle techniques the recorded images are only correlated and not reconstructed as for HI. Characteristic features of holography like the possibility to refocus on other object planes in the reconstruction process are not possible with speckle metrology.

Our idea was to transfer all methods of classical HI using photographic plates to Digital Holography. Surprisingly we discovered, that Digital Holography offers more possibilities than classical HI: The wavefronts can be manipulated in the numerical reconstruction process, enabling operations not possible in optical holography. Especially the interference phase can be calculated directly from the holograms, without evaluation of an interference pattern.

The efficiency of Digital Holography depends strongly on the resolution of the electronic target used to record the holograms. When we made our first experiments in the nineties of the last century Charged Coupled Devices begun to replace analogue sensors in cameras. The resolution of commercially available cameras was quite low, about some hundred pixels per line, and the output signal of cameras already equipped with CCD's was still analogue. In those days digital sampling of camera images and running of routines for numerical hologram reconstruction was only possible on special digital image processing hardware and not, as today, on ordinary PC's. The reconstruction of a hologram digitized with 512 × 512 pixels took about half an hour in 1991 on a Digital Image Processing unit developed at BIAS especially for optical metrology purposes. Nevertheless we made our first experiments with this type of cameras. Today numerical reconstruction of holograms with 1 million pixel is possible nearly in real time on state of the art PC's.

Then fully digital CCD cameras with 1 million pixels and smaller pixels than those of the previous camera generation emerged on the market. These cameras showed better performance and first applications in optical metrology became possible. Today digital CCD cameras with 4 million pixels are standard.

The tremendous development in opto-electronics and in data processing pushed Digital Holography to new perspectives: It is applied with success in optical deformation and strain analysis, shape measurement, microscopy and for investigations of flows in liquids and gases. In this book we make the trial to describe the principles of this method and to report on the various applications. We took pains to prepare the manuscript carefully and to avoid mistakes. However, we are not perfect. Comments, suggestions for improvements or corrections are therefore welcome and will be considered in potential further editions.

Some pictures in this book originate from common publications with other co-authors. All of our co-workers, especially W. Osten, Th. Kreis, D. Holstein, S. Seebacher, H.-J. Hartmann and V. Kebbel are gratefully acknowledged.

# Contents

# 1 Introduction

Dennis Gabor invented holography in 1948 as a method for recording and reconstructing amplitude and phase of a wave field [37, 38, 39]. He created the word *holography* from the Greek words 'holos' meaning whole or entire and 'graphein' meaning to write.

A holographically stored image or hologram is the photographically or otherwise recorded interference pattern between a wave field scattered from the object and a coherent background named reference wave. It is usually recorded on a flat surface, but contains the information about the entire three-dimensional wave field. This information is coded in form of interference stripes, usually not visible for the human eye due to the high spatial frequencies. The object wave can be reconstructed by illuminating the hologram with the reference wave again. This reconstructed wave is by passive means indistinguishable from the original object wave. An observer recognizes a three-dimensional image with all effects of perspective and depth of focus.

Gabor illuminated in his original set-up the hologram by a parallel beam through the mostly transparent object. Therefore the axes of both the object wave and the reference wave were parallel. The reconstruction of this hologram results in the real image superimposed by the undiffracted part of the reconstruction wave and the so called 'twin image' (or virtual image) laying on the optical axis, i.e. in-line. Significant improvements of this in-line holography were made by Leith and Upatnieks, who introduced an off-axis reference wave [91, 92]. Their set-up separates the two images and the reconstruction wave spatially.

One major application of holography is Holographic Interferometry (HI), developed in the late sixties of the last century by Stetson, Powell [127, 151] and others. HI made it possible to map the displacements of rough surfaces with an accuracy of a fraction of a micrometer. It also enabled interferometric comparisons of stored wave fronts existing at different times.

The development of computer technology allowed to transfer either the recording process or the reconstruction process into the computer. The first approach led to *Computer Generated Holography* (CGH), which generates artificial holograms by numerical methods. Afterwards these computer generated holograms are reconstructed optically. This technique is not considered here, the interested reader is referred to the literature, see e.g. Lee [90], Bryngdahl and Wyrowski [11] or Schreier [140].

Numerical hologram reconstruction was initiated by Goodman and Lawrence [42] and by Yaroslavskii, Merzlyakov and Kronrod [81]. They sampled optically enlarged parts of in-line and Fourier holograms recorded on a photographic plate.

These digitized conventional' holograms were reconstructed numerically. Onural and Scott improved the reconstruction algorithm and applied this method for particle measurement [106, 93, 105]. Haddad et al. described a holographic microscope based on numerical reconstruction of Fourier holograms [46].

A big step forward was the development of direct recording of Fresnel holograms with Charged Coupled Devices (CCD's) by Schnars and Jüptner [130, 131]. This method enables now full digital recording and processing of holograms, without any photographic recording as intermediate step. The name which has been originally proposed for this technique was 'direct holography' [130], emphasizing the direct way from optical recording to numerical processing. Later on the term **Digital Holography** has been accepted in the optical metrology community for this method. Although this name is sometimes also used for Computer Generated Holography, Digital Holography is used within the scope of this book as a designation for digital recording and numerical reconstruction of holograms.

Schnars and Jüptner applied Digital Holography to interferometry and demonstrated that digital hologram reconstruction offers much more possibilities than conventional (optical) processing: The phases of the stored light waves can be calculated directly from the digital holograms, without generating phase shifted interferograms [128, 129]. Other methods of optical metrology, such as shearography or speckle photography, can be derived numerically from Digital Holography [132]. Thus one can choose the interferometric technique (hologram interferometry, shearography or other techniques) after hologram recording by mathematical methods.

The use of electronic devices such as CCD's for recording of interferograms was already established in Electronic Speckle Pattern Interferometry (ESPI, also named *TV-holography*), discovered independently from each other by Butters and Leendertz [12], Macovski, Ramsey and Schaefer [96] and Schwomma [141]: Two speckle interferograms are recorded in different states of the object under investigation. The speckle patterns are subtracted electronically. The resulting fringe pattern has some similarities to that of conventional or digital HI. Main differences are the speckle appearance of the fringes and the loss of phase in the correlation process [29, 94, 95]. The interference phase has to be recovered with phase shifting methods [19, 152, 153], requiring additional experimental effort (phase shifting unit). Digital Holographic Interferometry and ESPI are competing methods: The image subtraction in ESPI is easier than the numerical reconstruction of Digital Holography, but the information content of digital holograms is higher. ESPI and other methods of speckle metrology are also discussed in this book in order to compare them with Digital Holographic Interferometry.

Since mid nineties of the last century Digital Holography has been extended, improved and applied to several measurement tasks. Important steps are:

- improvements of the experimental techniques and of the reconstruction algorithm [21, 23, 24, 27, 28, 33, 43, 44, 66, 68, 72-76, 85, 98, 104, 116, 120, 121, 124, 134-136, 139, 148, 167, 175],
- applications in deformation analysis and shape measurement [18, 59, 70, 109, 110, 119, 133, 138, 142, 143, 163, 180],

- the development of phase shifting digital holography [30, 45, 54, 62, 86, 146, 165, 171-174, 178, 179],
- applications in imaging, particle tracking and microscopy [5, 6, 7, 22, 32, 55, 65, 67, 77, 114, 122, 123, 137, 158, 159, 160, 168, 169],
- measurement of refractive index distributions within transparent media due to temperature or concentration variations [31, 63, 64, 115, 166],
- applications in encrypting of information [56, 84, 155, 156],
- the development of digital light-in-flight holography and other short-coherence-length applications [13, 61, 101-103, 117, 126],
- the combination of digital holography with heterodyne techniques [88, 89]
- the development of methods to reconstruct the three-dimensional object structure from digital holograms [34, 35, 57, 97, 157, 177]
- the development of comparative Digital Holography [111, 112]
- the use of a Digital Mirror Device (DMD) for optical reconstruction of digital holograms (DMD) [78].

This book is structured as follows: Optical foundations and the basic principles of holography are discussed in chapter 2. The third chapter describes the process of digital hologram recording and the various methods of numerical reconstruction. Digital Holographic Interferometry is discussed in chapter 4. The fifth chapter is devoted to Digital Holographic Microscopy. Special applications and techniques like short coherence recording, particle field applications and comparative holography are described in chapter 6. In the last chapter 7 speckle metrology techniques are discussed and compared with Digital Holographic Interferometry.

# 2 Fundamental Principles of Holography

## 2.1 Light Waves

Light can be described as an electromagnetic wave or as a current of particles called photons. The model to be referred to depends on the experiment under investigation. Both models contradict each other, but are necessary to describe the full spectrum of light phenomena. Interaction of light with the atomic structure of matter is described by quantum optics, the theory dealing with photons. Refraction, diffraction and interference are perfectly described by the wave model, which is based on the theory of classical electromagnetism.

Interference and diffraction form the basis of holography. The appropriate theory is therefore the wave model. The oscillating quantities are the electric and the magnetic fields. The field amplitudes oscillate perpendicularly to the propagation direction of light and perpendicularly to each other, i.e. light waves are transverse phenomena. Light waves can be described either by the electrical or by the magnetic field.

Light propagation is described by the wave equation, which follows from Maxwell equations. The wave equation in vacuum is

$$\nabla^2 \vec{E} - \frac{1}{c^2}\frac{\partial^2 \vec{E}}{\partial t^2} = 0 \tag{2.1}$$

Here $\vec{E}$ is the electric field and $\nabla^2$ is the *Laplace operator* defined as

$$\nabla^2 = \frac{\partial^2}{\partial x^2} + \frac{\partial^2}{\partial y^2} + \frac{\partial^2}{\partial z^2} \tag{2.2}$$

$c$ is the speed of light in vacuum:

$$c = 299792458\, m/s \tag{2.3}$$

The electrical field $\vec{E}$ is a vector quantity, which means it could vibrate in any direction, which is perpendicular to the light propagation. However, in many applications the wave vibrates only in a single plane. Such light is called *linear polarized light*. In this case it is sufficient to consider the scalar wave equation

$$\frac{\partial^2 E}{\partial z^2} - \frac{1}{c^2}\frac{\partial^2 E}{\partial t^2} = 0 \tag{2.4}$$

where the light propagates in z-direction.

It could be easily verified that a linearly polarized, harmonic plane wave described by

$$E(x,y,z,t) = a\cos\left(\omega t - \vec{k}\vec{r} - \varphi_0\right) \tag{2.5}$$

is a solution of the wave equation.

*E(x,y,z,t)* is the modulus of the electrical field vector at the point with spatial vector $\vec{r} = (x,y,z)$ at the time $t$. The quantity $a$ is named *amplitude*. The *wave vector* $\vec{k}$ describes the propagation direction of the wave:

$$\vec{k} = k\vec{n} \tag{2.6}$$

$\vec{n}$ is a unit vector in propagation direction. Points of equal phase are located on parallel planes that are perpendicular to the propagation direction. The modulus of $\vec{k}$ named *wave number* is calculated by

$$\left|\vec{k}\right| \equiv k = \frac{2\pi}{\lambda} \tag{2.7}$$

The angular frequency $\omega$ corresponds to the frequency $f$ of the light wave by

$$\omega = 2\pi f \tag{2.8}$$

Frequency $f$ and wavelength $\lambda$ are related by the speed of light $c$:

$$c = \lambda f \tag{2.9}$$

The spatially varying term

$$\varphi = -\vec{k}\vec{r} - \varphi_0 \tag{2.10}$$

is named *phase*, with phase constant $\varphi_0$. It has to be pointed out that this definition is not standardized. Some authors designate the entire argument of the cosine function, $\omega t - \vec{k}\vec{r} - \varphi_0$, as phase. The definition Eq. (2.10) is favourable to describe the holographic process and therefore used in this book.

The vacuum wavelengths of visible light are in the range of *400 nm* (violet) to *780 nm* (deep red). The corresponding frequency range is $7.5\cdot 10^{14}\,Hz$ to $3.8\cdot 10^{14}\,Hz$. Light sensors as e. g. the human eye, photodiodes, photographic films or CCD's are not able to detect such high frequencies due to technical and physical reasons. The only directly measurable quantity is the *intensity*. It is proportional to the time average of the square of the electrical field:

$$I = \varepsilon_0 c\left\langle E^2\right\rangle_t = \varepsilon_0 c \lim_{T\to\infty}\frac{1}{2T}\int_{-T}^{T} E^2 dt \tag{2.11}$$

$\langle\ \rangle_t$ stands for the time average over many light periods. The constant factor $\varepsilon_0 c$ results if the intensity is formally derived Maxwell equations. The constant $\varepsilon_0$ is the vacuum permittivity.

For a plane wave Eq. (2.5) has to be inserted:

$$I = \varepsilon_0 c a^2 \left\langle \cos^2\left(\omega t - \vec{k}\vec{r} - \varphi_0\right)\right\rangle_t = \frac{1}{2}\varepsilon_0 c a^2 \tag{2.12}$$

According to Eq. (2.12) the intensity is calculated by the square of the amplitude.

The expression (2.5) can be written in complex form as

$$E(x,y,z,t) = a\,\mathrm{Re}\left\{\exp\left(i\left(\omega t - \vec{k}\vec{r} - \varphi_0\right)\right)\right\} \tag{2.13}$$

where 'Re' denotes the real part of the complex function. For computations this 'Re' can be omitted. However, only the real part represents the physical wave:

$$E(x,y,z,t) = a\exp\left(i\left(\omega t - \vec{k}\vec{r} - \varphi_0\right)\right) \tag{2.14}$$

One advantage of the complex representation is that the spatial and temporal parts factorize:

$$E(x,y,z,t) = a\exp(i\varphi)\exp(i\omega t) \tag{2.15}$$

In many calculations of optics only the spatial distribution of the wave is of interest. In this case only the spatial part of the electrical field, named *complex amplitude*, has to be considered:

$$A(x,y,z) = a\exp(i\varphi) \tag{2.16}$$

The equations (2.15) and (2.16) are not just valid for plane waves, but in general for three-dimensional waves whose amplitude $a$ and phase $\varphi$ may be functions of $x$, $y$ and $z$.

In complex notation the intensity is now simply calculated by taking the square of the modulus of the complex amplitude

$$I = \frac{1}{2}\varepsilon_0 c|A|^2 = \frac{1}{2}\varepsilon_0 c A^* A = \frac{1}{2}\varepsilon_0 c a^2 \tag{2.17}$$

where * denotes the conjugate complex. In many practical calculations where the absolute value of $I$ is not of interest the factor $1/2\varepsilon_0 c$ can be neglected, which means the intensity is simply calculated by $I = |A|^2$.

## 2.2 Interference

The superposition of two or more waves in space is named *interference*. If each single wave described by $\vec{E}_i(\vec{r},t)$ is a solution of the wave equation, also the superposition

$$\vec{E}(\vec{r},t)=\sum_i \vec{E}_i(\vec{r},t) \qquad i=1,2,... \tag{2.18}$$

is a solution, too. This is because the wave equation is a linear differential equation.

In the following interference of two monochromatic waves with equal frequencies and wavelengths is considered. The waves must have the same polarization directions, i. e. the scalar formalism can be used. The complex amplitudes of the waves are

$$A_1(x,y,z)=a_1\exp(i\varphi_1) \tag{2.19}$$

$$A_2(x,y,z)=a_2\exp(i\varphi_2) \tag{2.20}$$

The resulting complex amplitude is then calculated by the sum of the individual amplitudes:

$$A=A_1+A_2 \tag{2.21}$$

According to Eq. (2.17) the intensity becomes

$$\begin{aligned} I &= |A_1+A_2|^2=(A_1+A_2)(A_1+A_2)^* \\ &= a_1^{\ 2}+a_2^{\ 2}+2a_1a_2\cos(\varphi_1-\varphi_2) \\ &= I_1+I_2+2\sqrt{I_1I_2}\cos\Delta\varphi \end{aligned} \tag{2.22}$$

$I_1$, $I_2$ being the individual intensities and

$$\Delta\varphi=\varphi_1-\varphi_2 \tag{2.23}$$

The resulting intensity is the sum of the individual intensities *plus* the interference term $2\sqrt{I_1I_2}\cos\Delta\varphi$, which depends on the phase difference between the waves. The intensity reaches its maximum in all points to which applies

$$\Delta\varphi=2n\pi \qquad \text{for } n=0,1,2,... \tag{2.24}$$

This is called *constructive interference*. The intensity reaches its minimum where

$$\Delta\varphi=(2n+1)\pi \qquad \text{for } n=0,1,2,... \tag{2.25}$$

This is named *destructive interference*. The integer $n$ is the interference order. An interference pattern consists of dark and bright fringes as a result of constructive and destructive interference. The scalar theory applied here can also be used for waves with different polarization directions, if components of the electric field vector are considered.

The superposition of two plane waves which intersect under an angle $\theta$ with respect to each other result in an interference pattern with equidistant spacing, figure 2.1. The fringe spacing $d$ is the distance from one interference maximum to the next and can be calculated by geometrical considerations. Figure 2.1 shows evidently that

$$\sin\theta_1 = \frac{\Delta l_1}{d} \quad ; \quad \sin\theta_2 = \frac{\Delta l_2}{d} \tag{2.26}$$

The quantities $\theta_1$ and $\theta_2$ are the angles between the propagation directions of the wavefronts and the vertical of the screen. The length $\Delta l_2$ is the path difference of wavefront W2 with respect to wavefront W1 at the position of the interference maximum P1 (W2 has to travel a longer way to P1 than W1). At the neighboring maximum P2 the conditions are exchanged: Now W1 has to travel a longer way; the path difference of W2 with respect to W1 is $-\Delta l_1$. The variation between the path differences at neighboring maxima is therefore $\Delta l_1 + \Delta l_2$. This difference is equal to one wavelength. Thus the interference condition is:

$$\Delta l_1 + \Delta l_2 = \lambda \tag{2.27}$$

Combining Eq. (2.26) with (2.27) results to:

$$d = \frac{\lambda}{\sin\theta_1 + \sin\theta_2} = \frac{\lambda}{2\sin\frac{\theta_1+\theta_2}{2}\cos\frac{\theta_1-\theta_2}{2}} \tag{2.28}$$

The approximation $\cos(\theta_1 - \theta_2)/2 \approx 1$ and $\theta = \theta_1 + \theta_2$ gives:

$$d = \frac{\lambda}{2\sin\frac{\theta}{2}} \tag{2.29}$$

Instead of the fringe spacing $d$ the fringe pattern can also be described by the spatial frequency $f$, which is the reciprocal of $d$:

$$f = d^{-1} = \frac{2}{\lambda}\sin\frac{\theta}{2} \tag{2.30}$$

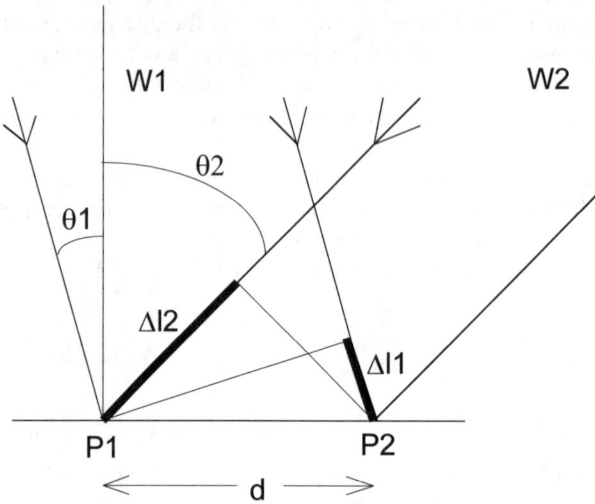


**Fig. 2.1.** Interference of two plane waves

## 2.3 Coherence

### 2.3.1 General

Generally the resulting intensity of two different sources, e. g. two electric light bulbs directed on a screen, is additive. Instead of dark and bright fringes as expected by Eq. (2.22) only a uniform brightness according to the sum of the individual intensities becomes visible.

In order to generate interfere fringes the phases of the individual waves must be correlated in a special way . This correlation property is named *coherence* and is investigated in this chapter. Coherence is the ability of light to interfere. The two aspects of coherence are the temporal and the spatial coherence. Temporal coherence describes the correlation of a wave with itself at different instants [71]. Spatial coherence depicts the mutual correlation of different parts of the same wavefront.

### 2.3.2 Temporal Coherence

The prototype of a two beam interferometer is the Michelson-interferometer, see figure 2.2. Light emitted by the light source S is split into two partial waves by the beam splitter BS. The partial waves travel to the mirrors M1 respectively M2, and are reflected back into the incident directions. After passing the beam splitter again they are superimposed at a screen. Usually the superimposed partial waves are not exactly parallel, but are interfering with a small angle. As a result a two-dimensional interference pattern becomes visible.

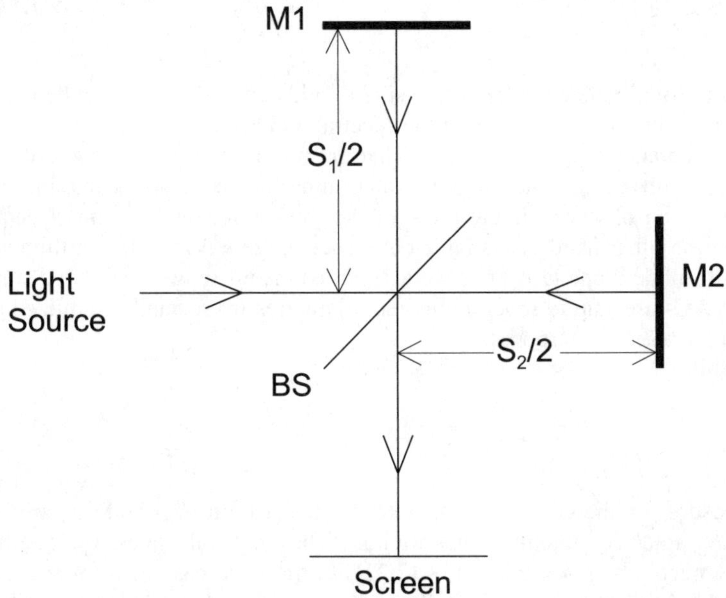


**Fig. 2.2.** Michelson's interferometer

The optical path length from BS to M1 and back to BS is $s_1$, the optical path length from BS to M2 and back to BS is $s_2$. Experiments prove that interference can only occur if the optical path difference $s_1 - s_2$ does not exceed a certain length $L$. If the optical path difference exceeds this limit, the interference fringes vanish and just an uniform brightness becomes visible on the screen. The qualitative explanation for this phenomenon is as follows: Interference fringes can only develop if the superimposed waves have a well defined (constant) phase relation. The phase difference between waves emitted by different sources vary randomly and thus the waves do not interfere. The atoms of the light source emit wave trains with finite length $L$. If the optical path difference exceeds this wave train length, the partial waves belonging together do not overlap after passing the different ways and interference is not possible.

The critical path length difference or, equivalently, the length of a wave train is named *coherence length L.* The corresponding emission time for the wave train

$$\tau = \frac{L}{c} \tag{2.31}$$

is called *coherence time.*

According to the laws of Fourier analysis a wave train with finite length $L$ corresponds to light with finite spectral width $\Delta f$:

$$L = \frac{c}{\Delta f} \tag{2.32}$$

Light with long coherence length is called highly monochromatic. The coherence length is therefore a measure for the spectral width.

Typical coherence lengths of light radiated from thermal sources, e. g. ordinary electric light bulbs, are in the range of some micrometers. That means, interference can only be observed if the arms of the interferometer have nearly equal lengths. On the other hand lasers have coherence lengths from a few millimetres (e. g. a multi-mode diode laser) to several hundred meters (e. g. a stabilized single mode Nd:YAG-laser) up to several hundred kilometres for specially stabilized gas lasers used for research purposes.

The visibility

$$V = \frac{I_{\max} - I_{\min}}{I_{\max} + I_{\min}} \tag{2.33}$$

is a measure for the contrast of an interference pattern. $I_{max}$ and $I_{min}$ are two neighbouring intensity maxima and minima. They are calculated by inserting $\Delta\phi = 0$, respectively $\Delta\phi = \pi$ into Eq. (2.22). In the ideal case of infinite coherence length the visibility is therefore

$$V = \frac{2\sqrt{I_1 I_2}}{I_1 + I_2} \tag{2.34}$$

To consider the effect of finite coherence length the *complex self coherence* $\Gamma(\tau)$ is introduced:

$$\begin{aligned} \Gamma(\tau) &= \langle E(t+\tau)E^*(t)\rangle \\ &= \lim_{T\to\infty} \frac{1}{2T} \int_{T}^{T} E(t+\tau)E^*(t)dt \end{aligned} \tag{2.35}$$

$E(t)$ is the electrical field (to be precise: the complex analytical signal) of one partial wave while $E(t+\tau)$ is the electrical field of the other partial wave. The latter is delayed in time by $\tau$. Eq. (2.35) is the autocorrelation of $E$. The normalized quantity

$$\gamma(\tau) = \frac{\Gamma(\tau)}{\Gamma(0)} \tag{2.36}$$

defines the degree of coherence.

With finite coherence length the interference equation (2.22) has to be replaced by

$$I = I_1 + I_2 + 2\sqrt{I_1 I_2}\,|\gamma|\cos\Delta\varphi \tag{2.37}$$

Maximum and minimum intensity are now calculated by

$$\begin{aligned} I_{\max} &= I_1 + I_2 + 2\sqrt{I_1 I_2}\,|\gamma| \\ I_{\min} &= I_1 + I_2 - 2\sqrt{I_1 I_2}\,|\gamma| \end{aligned} \tag{2.38}$$

Inserting these quantities into Eq. (2.33) gives

$$V = \frac{2\sqrt{I_1 I_2}}{I_1 + I_2}|\gamma| \tag{2.39}$$

For two partial waves with the same intensity, $I_1 = I_2$, Eq. (2.39) becomes

$$V = |\gamma| \tag{2.40}$$

$|\gamma|$ is equal to the visibility and therefore a measure of the ability of the two wave fields to interfere. $|\gamma| = 1$ describes ideally monochromatic light or, likewise, light with infinite coherence length. $|\gamma| = 0$ is true for completely incoherent light. Partially coherent light is described by $0 < |\gamma| < 1$.

### 2.3.3 Spatial Coherence

Spatial coherence describes the mutual correlation of different parts of the same wavefront. This property is measured with the Young interferometer, figure 2.3. An extended light source emits light from different points. Possible interferences are observed on a screen. An aperture with two transparent holes is mounted between light source and screen. Under certain conditions, which will be derived in this chapter, interferences are visible on the screen. The fringes result from light rays which travelled on different ways to the screen, either via the upper or via the lower hole in the aperture [164]. The interference pattern vanishes if the distance between the holes $a$ exceeds the critical limit $a_k$. This limit is named *coherence distance*. The phenomenon is not related to the spectral width of the light source, but has following cause: The waves emitted by different source points of the extended light source are superimposed on the screen. It may happen, that a special source point generates an interference maximum at a certain point on the screen, while another source point generates a minimum at the same screen point. This is because the optical path length is different for light rays emerging from different source points. In general the contributions from all source points compensate themselves and the contrast vanishes. This compensation is avoided if following condition is fulfilled for every point of the light source:

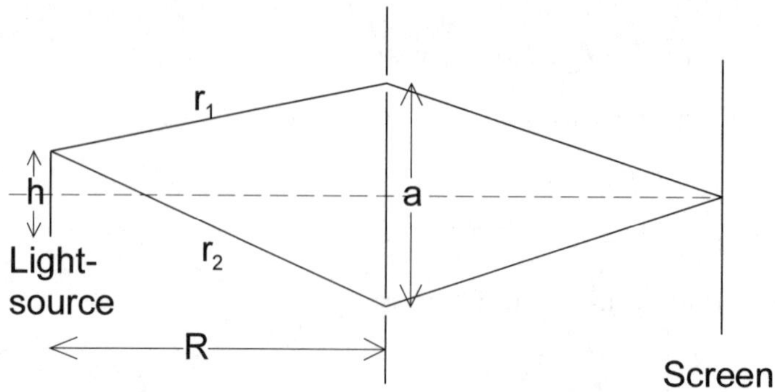


**Fig. 2.3.** Young's interferometer

$$r_2 - r_1 < \frac{\lambda}{2} \tag{2.41}$$

This condition is definitely fulfilled, if it is kept for the edges of the light source. The following relations are valid for points at the edges:

$$r_1^2 = R^2 + \left(\frac{a-h}{2}\right)^2 \quad ; \quad r_2^2 = R^2 + \left(\frac{a+h}{2}\right)^2 \tag{2.42}$$

$h$ is the width of the light source. Using of the assumptions $a \ll R$ and $h \ll R$ results to

$$r_2 - r_1 \approx \frac{ah}{2R} \tag{2.43}$$

Combining Eq. (2.41) and (2.43) leads to following expression:

$$\frac{ah}{2R} < \frac{\lambda}{2} \tag{2.44}$$

The coherence distance is therefore:

$$\frac{a_k h}{2R} = \frac{\lambda}{2} \tag{2.45}$$

In contrast to temporal coherence the spatial coherence depends not only on properties of the light source, but also on the geometry of the interferometer. A light source may initially generate interference, which means Eq. (2.44) is fulfilled. If the distance between the holes increases or the distance between the light source and the aperture decreases, Eq. (2.44) becomes violated and the interference figure vanishes.

To consider spatial coherence the autocorrelation function defined in Eq. (2.35) is extended:

$$\Gamma(\vec{r}_1,\vec{r}_2,\tau)=\left\langle E(\vec{r}_1,t+\tau)E^*(\vec{r}_2,t)\right\rangle \tag{2.46}$$

$$=\lim_{T\to\infty}\frac{1}{2T}\int_{-T}^{T}E(\vec{r}_1,t+\tau)E^*(\vec{r}_2,t)dt$$

$\vec{r}_1$, $\vec{r}_2$ are the spatial vectors of the holes in the aperture of the Young interferometer. This function is named cross correlation function. The normalized function is

$$\gamma(\vec{r}_1,\vec{r}_2,\tau)=\frac{\Gamma(\vec{r}_1,\vec{r}_2,\tau)}{\sqrt{\Gamma(\vec{r}_1,\vec{r}_1,0)\Gamma(\vec{r}_2,\vec{r}_2,0)}} \tag{2.47}$$

where $\Gamma(\vec{r}_1,\vec{r}_1,0)$ is the intensity at $\vec{r}_1$ and $\Gamma(\vec{r}_2,\vec{r}_2,0)$ is the intensity at $\vec{r}_2$. Eq. (2.47) describes the degree of correlation between the light field at $\vec{r}_1$ at time $t+\tau$ with the light field at $\vec{r}_2$ at time *t*. The special function $\gamma(\vec{r}_1,\vec{r}_2,\tau=0)$ is a measure for the correlation between the field amplitudes at $\vec{r}_1$ and $\vec{r}_2$ at the same time and is named *complex degree of coherence*. The modulus of the normalized coherence function $|\gamma(\vec{r}_1,\vec{r}_2,\tau)|$ is measured with the Young interferometer.

## 2.4 Diffraction

A light wave which hits an obstacle is considered. This might be an opaque screen with some transparent holes, or vice versa, a transparent medium with opaque structures. From geometrical optics it is known that the shadow becomes visible on a screen behind the obstacle. By closer examination, one finds that this is not strictly correct. If the dimensions of the obstacle (e. g. diameter of holes in an opaque screen or size of opaque particles in a transparent volume) are in the range of the wavelength, the light distribution is not sharply bounded, but forms a pattern of dark and bright regions. This phenomenon is named *diffraction*.

Diffraction can be explained qualitatively with the *Huygens' principle*:

*Every point of a wavefront can be considered as a source point for secondary spherical waves. The wavefront at any other place is the coherent superposition of these secondary waves.*

Huygens' principle is graphically explained in figure 2.4.

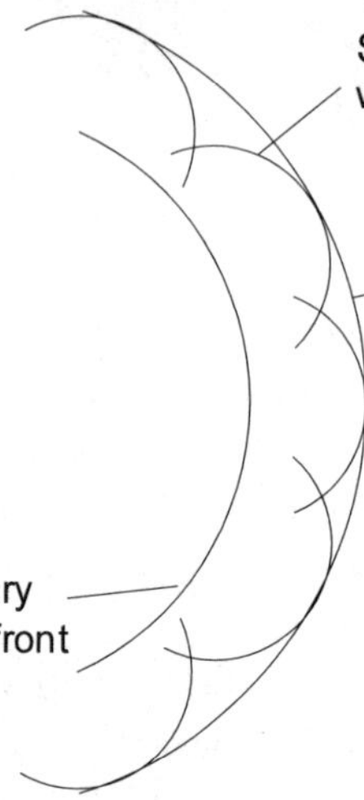


**Fig. 2.4.** Huygens' principle

The Fresnel-Kirchhoff integral describes diffraction quantitatively [69]:

$$\Gamma(\xi',\eta') = \frac{i}{\lambda}\int_{-\infty}^{\infty}\int_{-\infty}^{\infty} A(x,y)\frac{\exp\left(-i\frac{2\pi}{\lambda}\rho'\right)}{\rho'}Q\,dx\,dy \tag{2.48}$$

with

$$\rho' = \sqrt{(x-\xi')^2 + (y-\eta')^2 + d^2} \tag{2.49}$$

and

$$Q = \frac{1}{2}(\cos\theta + \cos\theta') \tag{2.50}$$

$A(x,y)$ is the complex amplitude in the plane of the bending aperture, see coordinate system defined in figure 2.5. $\Gamma(\xi',\eta')$ is the field in the observation plane. $\rho'$ stands for the distance between a point in the aperture plane and a point in the observation plane. Eq. (2.48) can be understood as the mathematical formulation of Huygens' principle:

The light source S lying in the source plane with coordinates $(\xi,\eta)$ radiates spherical waves. $A(x,y)$ is the complex amplitude of such a wave in the aperture plane. At first an opaque aperture with only one hole at the position $(x,y)$ is considered. Such a hole is now the source for secondary waves. The field at the position $(\xi',\eta')$ of the diffraction plane is proportional to the field at the entrance side of the aperture $A(x,y)$ and to the field of the secondary spherical wave emerging from (*x,y*), described by $\exp(-i2\pi/\lambda\,\rho')/\rho'$. Now the entire aperture as a plane consisting of many sources for secondary waves is considered. The entire resulting field in the diffraction plane is therefore the integral over all secondary spherical waves, emerging from the aperture plane.

The Huygens' principle would allow that the secondary waves not only propagate in the forward direction, but also back into the direction of the source. Yet, the experiment demonstrates that the wavefronts always propagate in one direction. To exclude this unphysical situation formally the inclination factor $Q$ defined in Eq. (2.50) is introduced in the Fresnel-Kirchhoff integral. $Q$ depends on the angle $\theta$ between the incident ray from the source and the unit vector $\vec{n}$ perpendicular to the aperture plane, and on the angle θ' between the bended ray and $\vec{n}$, see figure 2.6. $Q$ is approximately zero for $\theta \approx 0$ and $\theta' \approx \pi$. This prevents waves travelling into the backward direction. In most practical situations both θ and θ' are very small and $Q \approx 1$. The inclination factor can be considered as an ad hoc correction to the diffraction integral, as done here, or be derived in the formal diffraction theory [69, 41].

Other authors use a "+" sign in the argument of the exponential function of the Fresnel-Kirchhoff integral ($\Gamma(\xi,\eta) = ...A(x,y)\exp(+i2\pi/\lambda\,\rho')/\rho'...$) instead of the "-" sign used here. This depends on the definition of a harmonic wave in Eq. (2.14), which can be defined either as $\exp(+i\varphi)$ or $\exp(-i\varphi)$. However, using the "+" sign in Eq. (2.48) leads to the same expressions for all measurable quantities, as e.g. the intensity and the magnitude of the interference phase used in Digital Holographic Interferometry.

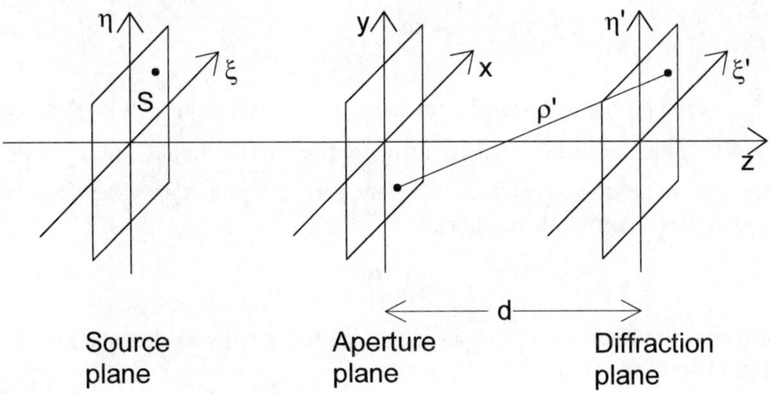


**Fig. 2.5.** Coordinate system

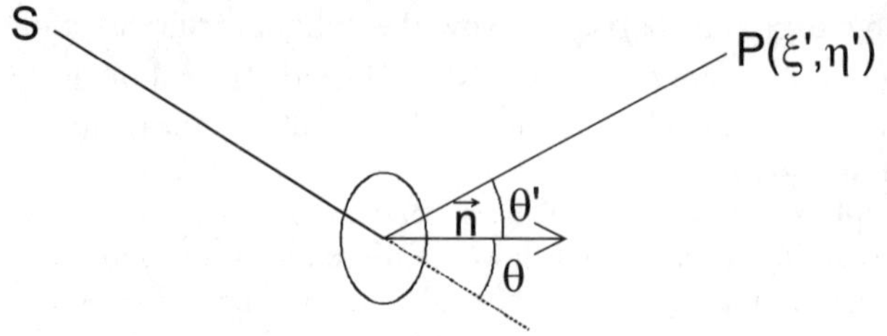


**Fig. 2.6.** Angles

## 2.5 Speckles

A rough surface illuminated with coherent light appears grainy for an observer. The intensity of the light scattered by the surface fluctuates randomly in space, dark and bright spots appear. These spots are named *speckles*, forming the entire image called speckle pattern, figure 2.7. A speckle pattern develops if the height variations of the rough surface are larger than the wavelength of the light.

Speckles result from interference of light scattered by the surface points. The phases of the waves scattered by different surface points fluctuate statistically due to the height variations. If these waves interfere with each other, a stationary speckle pattern develops.

It can be shown that the probability density function for the intensity in a speckle pattern obeys negative exponential statistics [40]:

$$P(I)dI = \frac{1}{\langle I \rangle} \exp\left(-\frac{I}{\langle I \rangle}\right) \tag{2.51}$$

$P(I)dI$ is the probability that the intensity at a certain point is lying between $I$ and $I + dI$. $\langle I \rangle$ is the mean intensity of the entire speckle field. The most probable intensity value of a speckle is therefore zero, i. e. most speckles are black. The standard deviation $\sigma_I$ is calculated by

$$\sigma_I = \langle I \rangle \tag{2.52}$$

That means the intensity variations are in the same order as the mean value. A usual definition of the contrast is

$$V = \frac{\sigma_I}{\langle I \rangle} \tag{2.53}$$

The contrast of a speckle pattern is therefore always unity.

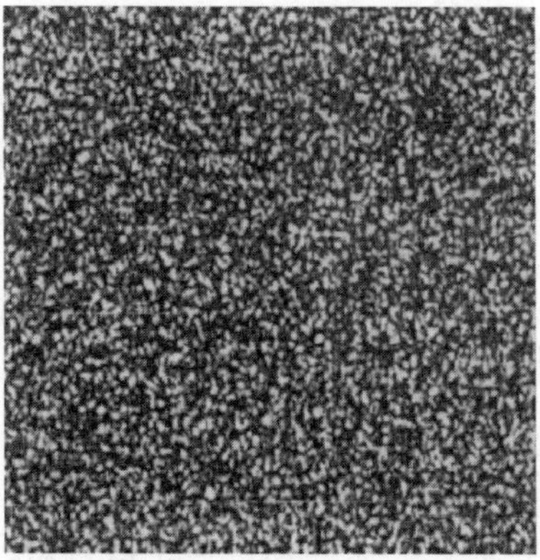

**Fig. 2.7.** Speckle pattern

One can distinguish between objective and subjective speckle formation. An objective speckle pattern develops on a screen, located in a distance $z$ from the illuminated surface, figure 2.8. There is no imaging system between surface and screen. The size of a speckle in an objective speckle pattern can be estimated using the spatial frequency formula Eq. (2.30). The two edge points of the illuminated surface form the highest spatial frequency:

$$f_{\max} = \frac{2}{\lambda}\sin\frac{\theta_{\max}}{2} \approx \frac{L}{\lambda z} \tag{2.54}$$

The reciprocal of $f_{max}$ is a measure for the speckle size:

$$d_{Sp} = \frac{\lambda z}{L} \tag{2.55}$$

A subjective speckle pattern develops if the illuminated surface is focussed with an imaging system, e. g. a camera lens or the human eye, figure 2.9. In this case the speckle diameter depends on the aperture diameter $a$ of the imaging system. The size of a speckle in a subjective speckle pattern can be estimated again using the spatial frequency:

$$f_{\max} = \frac{2}{\lambda}\sin\left(\frac{\theta_{\max}}{2}\right) \approx \frac{a}{\lambda b} \tag{2.56}$$

$b$ is the image distance of the imaging system. It follows for the speckle size:

$$d_{Sp} = \frac{\lambda b}{a} \tag{2.57}$$

The speckle size can be increased by closing the aperture of the imaging system.

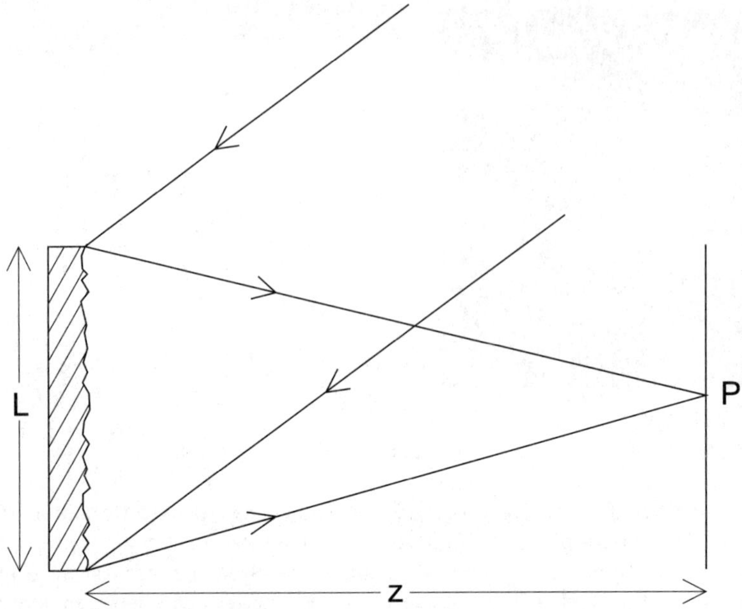


**Fig. 2.8.** Objective speckle formation

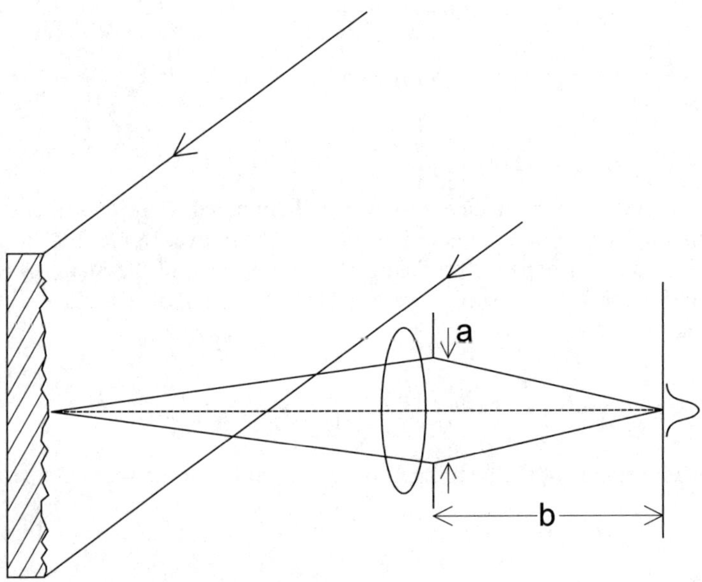


**Fig. 2.9.** Subjective speckle formation

## 2.6 Holography

### 2.6.1 Hologram Recording and Reconstruction

Holograms are usually recorded with an optical set-up consisting of a light source (laser), mirrors and lenses for beam guiding and a recording device, e. g. a photographic plate. A typical set-up is shown in figure 2.10 [47, 71]. Light with sufficient coherence length is split into two partial waves by a beam splitter (BS). The first wave illuminates the object. It is scattered at the object surface and reflected to the recording medium. The second wave - named reference wave - illuminates the light sensitive medium directly. Both waves interfere. The interference pattern is recorded, e.g. by chemical development of the photographic plate. The recorded interference pattern is named hologram.

The original object wave is reconstructed by illuminating the hologram with the reference wave, figure 2.11. An observer sees a virtual image, which is indistinguishable from the original object. The reconstructed image exhibits all effects of perspective and depth of focus.

The holographic process is described mathematically using the formalism of chapter 2.2. The complex amplitude of the object wave is described by

$$E_O(x,y) = a_O(x,y)\exp(i\varphi_O(x,y)) \tag{2.58}$$

with real amplitude $a_O$ and phase $\varphi_O$.

$$E_R(x,y) = a_R(x,y)\exp(i\varphi_R(x,y)) \tag{2.59}$$

is the complex amplitude of the reference wave with real amplitude $a_R$ and phase $\varphi_R$.

Both waves interfere at the surface of the recording medium. The intensity is calculated by

$$\begin{aligned} I(x,y) &= |E_O(x,y) + E_R(x,y)|^2 \\ &= (E_O(x,y) + E_R(x,y))(E_O(x,y) + E_R(x,y))^* \\ &= E_R(x,y)E_R^*(x,y) + E_O(x,y)E_O^*(x,y) + E_O(x,y)E_R^*(x,y) + E_R(x,y)E_O^*(x,y) \end{aligned} \tag{2.60}$$

The amplitude transmission $h(x,y)$ of the developed photographic plate (or of other recording media) is proportional to $I(x,y)$:

$$h(x,y) = h_0 + \beta\tau I(x,y) \tag{2.61}$$

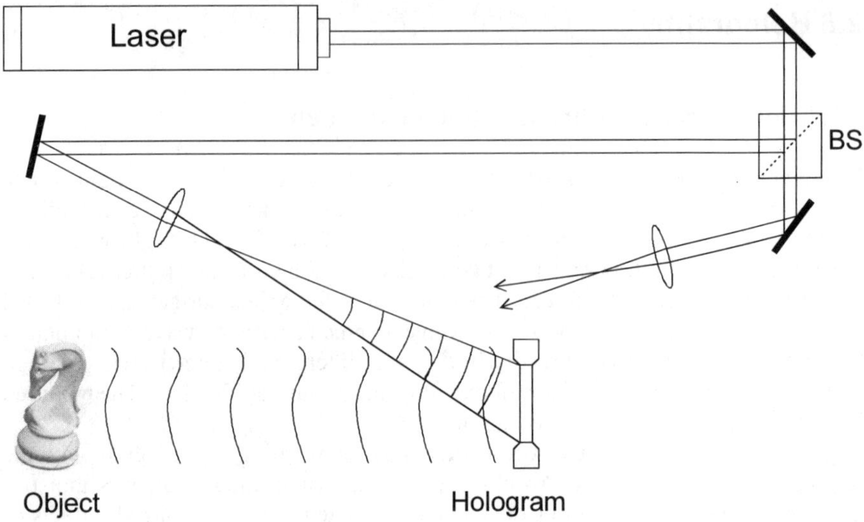


**Fig. 2.10.** Hologram recording

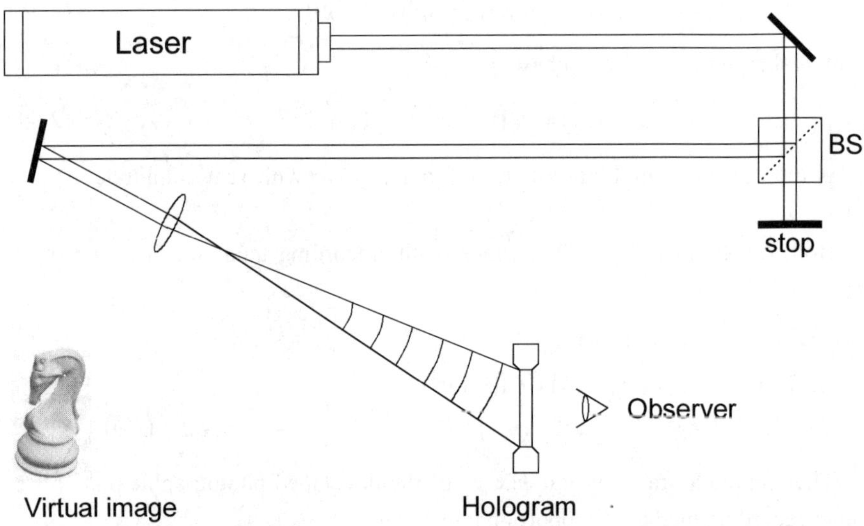


**Fig. 2.11.** Hologram reconstruction

The constant $\beta$ is the slope of the amplitude transmittance versus exposure characteristic of the light sensitive material. For photographic emulsions $\beta$ is negative. $\tau$ is the exposure time and $h_0$ is the amplitude transmission of the unexposed plate. *h(x,y)* is named hologram function. In Digital Holography using CCD's as recording medium $h_0$ can be neglected.

For hologram reconstruction the amplitude transmission has to be multiplied with the complex amplitude of the reconstruction (reference) wave:

$$E_R(x,y)h(x,y) = \left[h_0 + \beta\tau\left(a_R^2 + a_O^2\right)\right]E_R(x,y) + \beta\tau a_R^2 E_O(x,y) + \beta\tau E_R^2(x,y)E_O^*(x,y) \tag{2.62}$$

The first term on the right side of this equation is the reference wave, multiplied by a factor. It represents the undiffracted wave passing the hologram (zero diffraction order). The second term is the reconstructed object wave, forming the virtual image. The real factor $\beta\tau a_R^2$ only influences the brightness of the image. The third term generates a distorted real image of the object. For off-axis holography the virtual image, the real image and the undiffracted wave are spatially separated.

The reason for the distortion of the real image is the spatially varying complex factor $E_R^2$, which modulates the image forming conjugate object wave $E_O^*$. An undistorted real image can be generated by using the conjugate reference beam $E_R^*$ for reconstruction:

$$E_R^*(x,y)h(x,y) = \left[h_0 + \beta\tau\left(a_R^2 + a_O^2\right)\right]E_R^*(x,y) + \beta\tau a_R^2 E_O^*(x,y) + \beta\tau E_R^{*2}(x,y)E_O(x,y) \tag{2.63}$$

## 2.6.2 The Imaging Equations

The virtual image appears at the position of the original object if the hologram is reconstructed with the same parameters like those used in the recording process. However, if one changes the wavelength or the coordinates of the reconstruction wave source point with respect to the coordinates of the reference wave source point used in the recording process, the position of the reconstructed image moves. The coordinate shift is different for all points, thus the shape of the reconstructed object is distorted. The image magnification can be influenced by the reconstruction parameters, too.

The *imaging equations* relate the coordinates of an object point O with that of the corresponding point in the reconstructed image. These equations are quoted here without derivation, because they are needed to explain specific techniques such as Digital Holographic Microscopy. An exact derivation can be found in [47, 71].

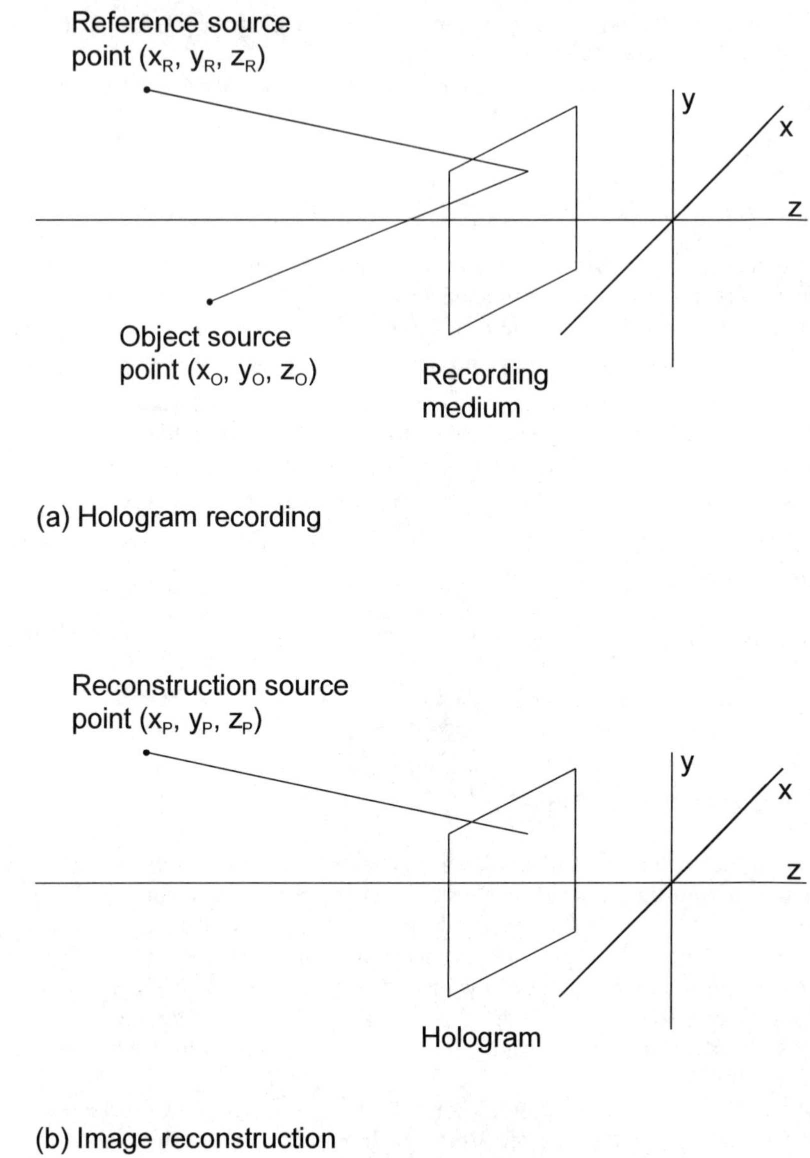


**Fig. 2.12.** Coordinate system used to describe holographic reconstruction

The coordinate system is shown in figure 2.12. $(x_O, y_O, z_O)$ are the coordinates of the object point O, $(x_R, y_R, z_R)$ are the coordinates of the source point of the reference wave used for hologram recording and $(x_P, y_P, z_P)$ are the coordinates of the source point of the reconstruction wave. $\mu = \lambda_2/\lambda_1$ denotes the ratio between the recording wavelength $\lambda_1$ and the reconstruction wavelength $\lambda_2$. The coordinates of that point in the reconstructed virtual image, which corresponds to the object point O, are:

$$x_1 = \frac{x_P z_O z_R + \mu x_O z_P z_R - \mu x_R z_P z_O}{z_O z_R + \mu z_P z_R - \mu z_P z_O} \tag{2.64}$$

$$y_1 = \frac{y_P z_O z_R + \mu y_O z_P z_R - \mu y_R z_P z_O}{z_O z_R + \mu z_P z_R - \mu z_P z_O} \tag{2.65}$$

$$z_1 = \frac{z_P z_O z_R}{z_O z_R + \mu z_P z_R - \mu z_P z_O} \tag{2.66}$$

The coordinates of that point in the reconstructed real image, which corresponds to the object point O, are:

$$x_2 = \frac{x_P z_O z_R - \mu x_O z_P z_R + \mu x_R z_P z_O}{z_O z_R - \mu z_P z_R + \mu z_P z_O} \tag{2.67}$$

$$y_2 = \frac{y_P z_O z_R - \mu y_O z_P z_R + \mu y_R z_P z_O}{z_O z_R - \mu z_P z_R + \mu z_P z_O} \tag{2.68}$$

$$z_2 = \frac{z_P z_O z_R}{z_O z_R - \mu z_P z_R + \mu z_P z_O} \tag{2.69}$$

An extended object can be considered to be made up of a number of point objects. The coordinates of all surface points are described by the above mentioned equations. The lateral magnification of the entire virtual image is depicted:

$$M_{lat,1} = \frac{dx_1}{dx_O} = \left[1 + z_0\left(\frac{1}{\mu z_P} - \frac{1}{z_R}\right)\right]^{-1} \tag{2.70}$$

The lateral magnification of the real image results in:

$$M_{lat,2} = \frac{dx_2}{dx_O} = \left[1 - z_0\left(\frac{1}{\mu z_P} + \frac{1}{z_R}\right)\right]^{-1} \tag{2.71}$$

The longitudinal magnification of the virtual image is given by:

$$M_{long,1} = \frac{dz_1}{dz_O} = \frac{1}{\mu} M_{lat,1}{}^2 \tag{2.72}$$

The longitudinal magnification of the real image is:

$$M_{long,2} = \frac{dz_2}{dz_O} = -\frac{1}{\mu} M_{lat,2}^{\;2} \tag{2.73}$$

There is a difference between real and virtual image to be mentioned: Since the real image is formed by the conjugate object wave $O^*$, it has the curious property that its depth is inverted. Corresponding points of the virtual image (which coincide with the original object points) and of the real image are located at equal distances from the hologram plane, but at opposite sides of it. The background and the foreground of the real image are therefore exchanged. The real image appears with the "wrong perspective". It is called a *pseudoscopic image*, in contrast to a normal or *orthoscopic* image.

## 2.7 Holographic Interferometry

### 2.7.1 Generation of Holographic Interferograms

Holographic Interferometry (HI) is a method to measure optical path length variations, which are caused by deformations of opaque bodies or refractive index variations in transparent media, e.g. fluids or gases [113]. HI is a non-contact, non-destructive method with very high sensitivity. Optical path changes up to one hundredth of a wavelength are resolvable.

Two coherent wave fields, which are reflected in two different states of the object, interfere. This is achieved e.g. in double-exposure holography by the recording of two wave fields on a single photographic plate, figure 2.13. The first exposure represents the object in its reference state, the second exposure represents the object in its loaded (e. g. deformed) state. The hologram is reconstructed by illumination with the reference wave, figure 2.14. As a result of the superposition of two holographic recordings with *slightly* different object waves only one image superimposed by interference fringes is visible, see example in figure 2.15. From this holographic interferogram the observer can determine optical path changes due to the object deformation or other effects.

In the real time technique the hologram is replaced - after processing - in exactly the recording position. When it is illuminated with the reference wave, the reconstructed virtual image coincides with the object. Interference patterns caused by phase changes between the holographically reconstructed reference object wave and the actual object wave are observable in real time.

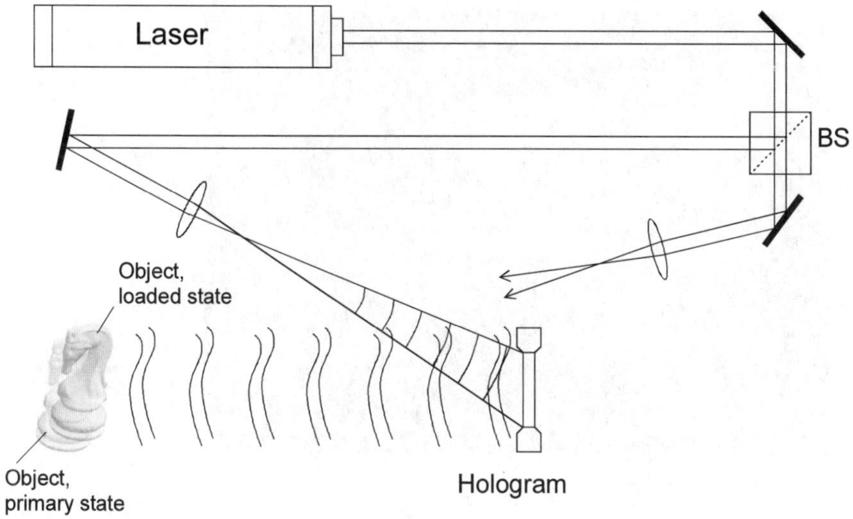


**Fig. 2.13.** Recording of a double exposed hologram

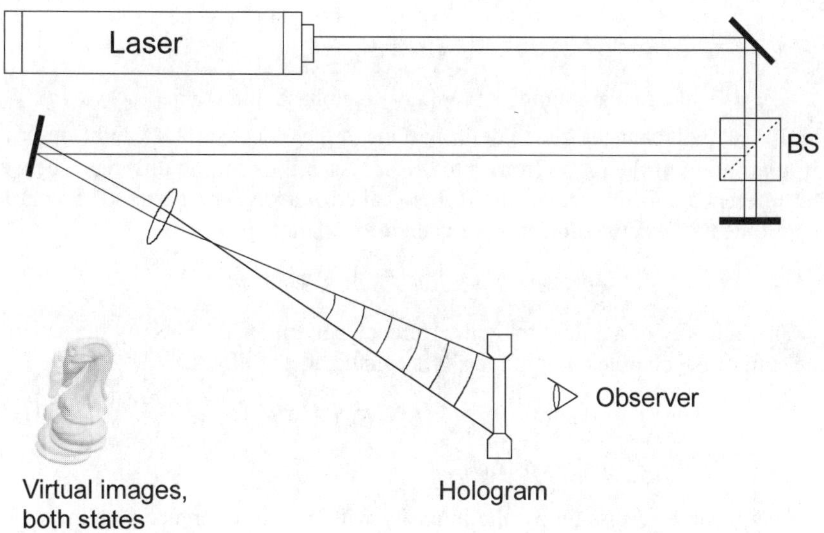


**Fig. 2.14.** Reconstruction

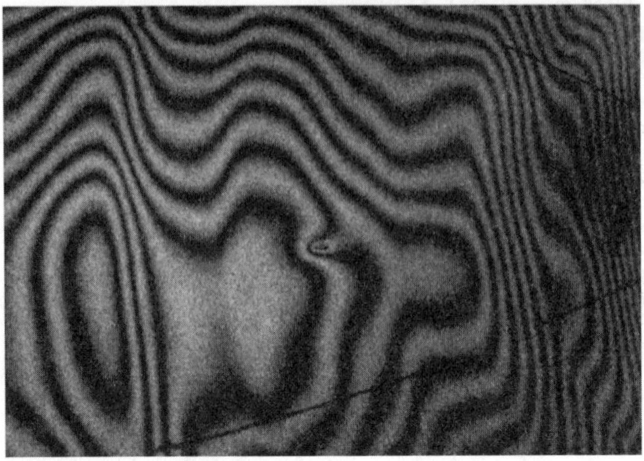

**Fig. 2.15.** Holographic interferogram

The following mathematical description is valid for the double exposure and for the real time technique. The complex amplitude of the object wave in the initial state is:

$$E_1(x,y) = a(x,y)\exp[i\varphi(x,y)] \tag{2.74}$$

$a(x,y)$ is the real amplitude and $\varphi(x,y)$ the phase of the object wave.

Optical path changes due to deformations of the object surface can be described by a variation of the phase from $\varphi$ to $\varphi+\Delta\varphi$. $\Delta\varphi$ stands for the difference between the reference and the actual phase. It is called *interference phase*. The complex amplitude of the actual object wave is therefore denoted by

$$E_2(x,y) = a(x,y)\exp[i(\varphi(x,y) + \Delta\varphi(x,y))] \tag{2.75}$$

The intensity of a holographic interference pattern is described by the square of the sum of the complex amplitudes. It is calculated as follows:

$$\begin{aligned} I(x,y) &= |E_1 + E_2|^2 = (E_1 + E_2)(E_1 + E_2)^* \\ &= 2a^2(1 + \cos(\Delta\varphi)) \end{aligned} \tag{2.76}$$

The general expression for the intensity within an interference pattern is therefore:

$$I(x,y) = A(x,y) + B(x,y)\cos\Delta\varphi(x,y) \tag{2.77}$$

The parameters $A(x,y)$ and $B(x,y)$ depend on the coordinates in the interferogram.

In practice these parameters are not known due to several disturbing effects:

- The object is illuminated by an expanded laser beam having a gaussian profile. The brightness of the holographic interferogram varies accordingly.
- The interferogram is superimposed by a high frequency speckle noise.
- Dust particles in the optical path result in diffraction patterns.
- The surface of the object under investigation may have a varying reflectivity influencing the brightness and visibility of the interferogram.
- Electronic recording and transmission of holographic interferograms generates additional noise.

Eq. (2.77) describes the relation between the intensity of the interference pattern and the interference phase, which contains the information about the physical quantity to be measured (object displacement, refractive index change or object shape). In general it is not possible to calculate $\Delta\varphi$ directly from the measured intensity, because the parameters *A(x,y)* and *B(x,y)* are not known. In addition the cosine is an even function (cos30°=cos-30°) and the sign of $\Delta\varphi$ is not determined unambiguously. Therefore several techniques have been developed to determine the interference phase by recording additional information. The most common techniques are the various phase shifting methods, which are briefly discussed in chapter 2.7.5.

## 2.7.2 Displacement Measurement by HI

In this chapter a relation between the measured interference phase and the displacement of the object surface under investigation is derived [71, 147]. The geometric quantities are explained in figure 2.16. The vector $\vec{d}(x,y,z)$ is called displacement vector. It describes the shift of a surface point from its initial position $P_1$ to the new position $P_2$ due to deformation. $\vec{s}_1$ and $\vec{s}_2$ are unit vectors from the illumination source point S to $P_1$, resp. $P_2$. $\vec{b}_1$ and $\vec{b}_2$ are unit vectors from $P_1$ to the observation point B, resp. from $P_2$ to B. The optical path difference between a ray from S to B via $P_1$ and a ray from S to B via $P_2$ is:

$$\begin{aligned}\delta &= \overline{SP_1} + \overline{P_1B} - \left(\overline{SP_2} + \overline{P_2B}\right) \\ &= \vec{s}_1\,\overrightarrow{SP_1} + \vec{b}_1\,\overrightarrow{P_1B} - \vec{s}_2\,\overrightarrow{SP_2} - \vec{b}_2\,\overrightarrow{P_2B}\end{aligned} \tag{2.78}$$

The lengths $\overline{SP_{1/2}}$ and $\overline{P_{1/2}B}$ are in the range of meter, while $\left|\vec{d}\right|$ is in the range of several micrometers. The vectors $\vec{s}_1$ and $\vec{s}_2$ can therefore be replaced by a unit vector $\vec{s}$ pointing into the bisector of the angle spread by $\vec{s}_1$ and $\vec{s}_2$:

$$\vec{s}_1 = \vec{s}_2 = \vec{s} \tag{2.79}$$

$\vec{b}_1$ and $\vec{b}_2$ are accordingly replaced by a unit vector $\vec{b}$ pointing into the bisector of the angle spread by $\vec{b}_1$ and $\vec{b}_2$:

$$\vec{b}_1 = \vec{b}_2 = \vec{b} \tag{2.80}$$

The displacement vector $\vec{d}(x, y, z)$ is given by:

$$\vec{d} = \overrightarrow{P_1B} - \overrightarrow{P_2B} \tag{2.81}$$

and

$$\vec{d} = \overrightarrow{SP_2} - \overrightarrow{SP_1} \tag{2.82}$$

Inserting Eq. (2.79) to (2.82) into (2.78) gives:

$$\delta = \left(\vec{b} - \vec{s}\right)\vec{d} \tag{2.83}$$

The following expression results for the interference phase:

$$\Delta\varphi(x, y) = \frac{2\pi}{\lambda}\vec{d}(x, y, z)\left(\vec{b} - \vec{s}\right) = \vec{d}(x, y, z)\vec{S} \tag{2.84}$$

The vector

$$\vec{S} = \frac{2\pi}{\lambda}\left(\vec{b} - \vec{s}\right) \tag{2.85}$$

is called *sensitivity vector*. The sensitivity vector is only defined by the geometry of the holographic arrangement. It gives the direction in which the set-up has maximum sensitivity. At each point the projection of the displacement vector onto the sensitivity vector is measured. Eq. (2.84) is the basis of all quantitative measurements of the deformation of opaque bodies.

In the general case of a three dimensional deformation field Eq. (2.84) contains the three components of $\vec{d}$ as unknown parameters. Three interferograms of the same surface with linear independent sensitivity vectors are necessary to determine the displacement. In many practical cases not the three dimensional displacement field is of interest, but the deformation perpendicular to the surface. This *out-of-plane* deformation can be measured using an optimised set-up with parallel illumination and observation directions ($\vec{S} = 2\pi/\lambda(0,0,2)$). The component $d_z$ is then calculated from the interference phase by

$$d_z = \Delta\varphi\frac{\lambda}{4\pi} \tag{2.86}$$

A phase variation of $2\pi$ corresponds to a deformation of $\lambda/2$.

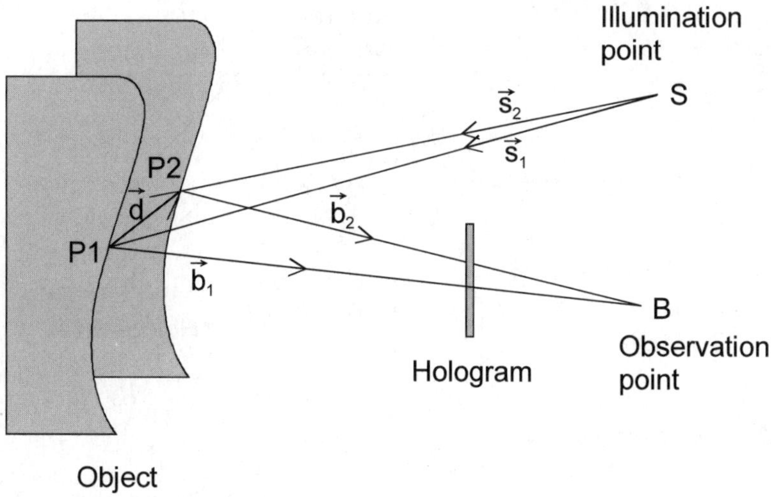


**Fig. 2.16.** Calculation of the interference phase

### 2.7.3 Holographic Contouring

Another application of HI is the generation of a fringe pattern corresponding to contours of constant elevation with respect to a reference plane. Such contour fringes can be used to determine the shape of a three-dimensional object.

Holographic contour interferograms can be generated by different methods. In the following the

- *two-wavelength method* and the
- *two-illumination-point method*

are described. A third method, *the two-refractive-index technique*, has less practical applications and is not considered here.

The principal set-up of the two-wavelength method is shown in figure 2.17. A plane wave illuminates the object surface . The back scattered light interferes with the plane reference wave at the holographic recording medium. In the set-up of figure 2.17 the illumination wave is reflected onto the object surface via a beam splitter in order to ensure parallel illumination and observation directions. Two holograms are recorded with different wavelengths $\lambda_1$ and $\lambda_2$ on the same photographic plate. This can be done either simultaneously using two lasers with different wavelengths or in succession changing the wavelength of a tuneable laser, e. g. a dye laser. After processing the double exposed hologram is replaced and reconstructed with only one of the two wavelengths, say $\lambda_2$. Two virtual images become visible. The image recorded with $\lambda_2$ coincides with the object surface. The other image, recorded with $\lambda_1$ but reconstructed with $\lambda_2$, is slightly distorted. The z-coordinate of this image $z'$ is calculated with the imaging equation (2.66):

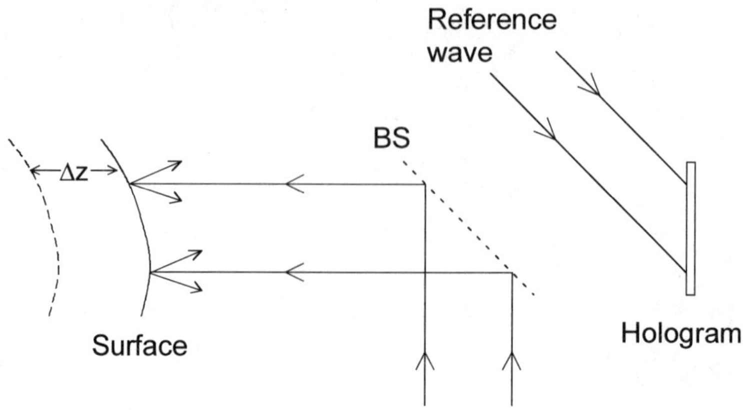


**Fig. 2.17.** Holographic contouring

$$z' = \frac{z_R^{\ 2} z}{z z_R + \frac{\lambda_2}{\lambda_1} z_R^{\ 2} - \frac{\lambda_2}{\lambda_1} z z_R} \approx z \frac{\lambda_1}{\lambda_2} \tag{2.87}$$

The indices "1" for virtual image ($z_1' \equiv z'$) and "O" for object ($z_O \equiv z$) are omitted and it is assumed not to change the source coordinates of the reconstruction wave with respect to those of the recording coordinates ($z_P \equiv z_R \to \infty$). The axial displacement of the image recorded with $\lambda_1$ but reconstructed with $\lambda_2$ with respect to the image recorded and reconstructed with $\lambda_2$ is therefore:

$$\Delta z = z' - z = z \frac{|\lambda_1 - \lambda_2|}{\lambda_2} \tag{2.88}$$

The path difference of the light rays on their way from the source to the surface and from the surface to the hologram is $2\Delta z$. The corresponding phase shift is:

$$\Delta\varphi(x, y) = \frac{2\pi}{\lambda_1} 2\Delta z = 4\pi z \frac{|\lambda_1 - \lambda_2|}{\lambda_1 \lambda_2} \tag{2.89}$$

The two shifted images interfere. According to Eq. (2.89) the phase shift depends on the distance $z$ from the hologram plane. All points of the object surface having the same z-coordinate (height) are therefore connected by a contour line. As a result an image of the surface superimposed by contour fringes develops.

The height jump between adjacent fringes is:

$$\Delta H = z(\Delta\varphi = (n+1)2\pi) - z(\Delta\varphi = n2\pi) = \frac{\lambda_1 \lambda_2}{2|\lambda_1 - \lambda_2|} = \frac{\Lambda}{2} \tag{2.90}$$

$\Lambda = \lambda_1\lambda_2 / |\lambda_1 - \lambda_2|$ is called *synthetic wavelength* or *equivalent wavelength*.

The object is intersected by parallel planes which have a distance of $\Delta H$ , see the principle in figure 2.18 and typical example in figure 2.19.

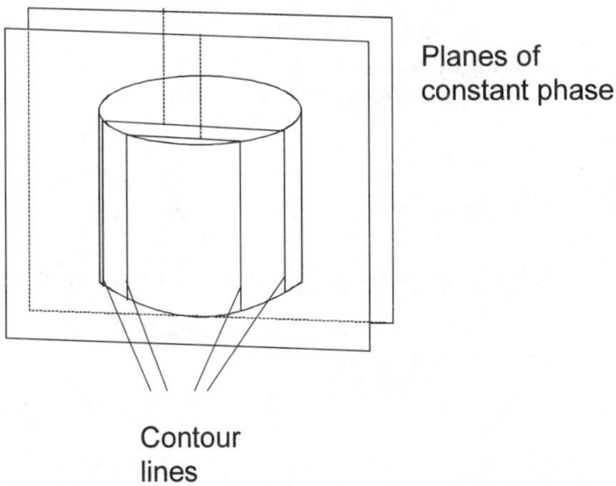


**Fig. 2.18.** Object intersection by contour lines

**Fig. 2.19.** Two-wavelength contour fringes

The equations derived in this chapter are valid only for small wavelength differences, because in addition to the axial displacement (which generates contour lines) also a lateral image displacement occurs. This lateral displacement can be neglected for small wavelength differences.

The principle of the two-illumination-point method is to make a double exposure hologram in which the point source illuminating the object is shifted slightly between the two exposures. If the illumination point S is shifted to S' between the two exposures (figure 2.20), the resulting optical path length difference $\delta$ is:

$$\begin{aligned}\delta &= \overline{SP} + \overline{PB} - \left(\overline{S'P} + \overline{PB}\right) = \overline{SP} - \overline{S'P} \\ &= \vec{s}_1 \overrightarrow{SP} - \vec{s}_2 \overrightarrow{S'P}\end{aligned} \tag{2.91}$$

The unit vectors $\vec{s}_1$ and $\vec{s}_2$ are defined as for the derivation of the interference phase due to deformation in chapter 2.7.2. The same approximation is used and these vectors are replaced by a common unit vector:

$$\vec{s}_1 = \vec{s}_2 = \vec{s} \tag{2.92}$$

Furthermore

$$\vec{p} = \overrightarrow{SP} - \overrightarrow{S'P} \tag{2.93}$$

is introduced as a vector from S to S'. The optical path difference is then given by

$$\delta = \vec{p}\,\vec{s} \tag{2.94}$$

The corresponding phase change is:

$$\Delta\varphi = \frac{2\pi}{\lambda}\vec{p}\,\vec{s} \tag{2.95}$$

The object surface is intersected by fringes which consist of a set of hyperboloids. Their common foci are the two points of illumination S and S'. If the dimensions of the object are small compared to the distances between the source points and the object, plane contouring surfaces result. A collimated illumination together with a telecentric imaging system also generates plane contouring surfaces. The distance between two neighbouring surfaces is

$$\Delta H = \frac{\lambda}{2\sin\dfrac{\theta}{2}} \tag{2.96}$$

where $\theta$ is the angle between the two illumination directions. Eq. (2.96) is analogue to the fringe spacing in an interference pattern formed by two intersecting plane waves, see Eq. (2.29) in chapter 2.2.

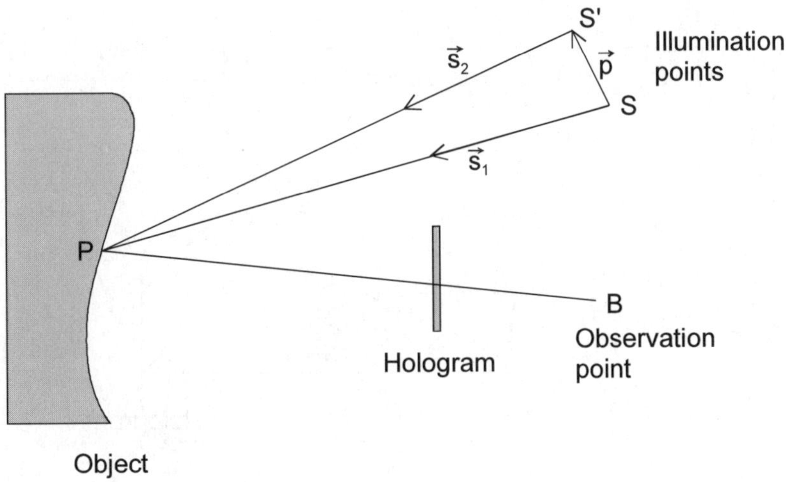


**Fig. 2.20.** Two-illumination point contouring

## 2.7.4 Refractive Index Measurement by HI

Another application of HI is the measurement of refractive index variations within transparent media. This mode of HI is used to determine temperature or concentration variations in fluid or gaseous media.

A refractive index change in a transparent medium causes a change of the optical path length and thereby a phase variation between two light waves passing the medium before and after the change. The interference phase due to refractive index variations is given by:

$$\Delta\varphi(x,y) = \frac{2\pi}{\lambda}\int_{l_1}^{l_2}\left[n(x,y,z) - n_0\right]dz \tag{2.97}$$

where $n_0$ is the refractive index of the medium under observation in its initial, unperturbed state and $n(x,y,z)$ is the final refractive index distribution. The light passes the medium in z-direction and the integration is taken along the propagation direction. Eq. (2.97) is valid for small refractive index gradients, where the light rays propagate along straight lines. The simplest case is that of a two-dimensional phase object with no variation of refractive index in z-direction. In this case the refractive index distribution $n(x,y)$ can be calculated directly from Eq. (2.97). In the general case of a refractive index varying also in z-direction Eq. (2.97) cannot be solved without further information about the process. However, in many practical experiments only two-dimensional phase objects have to be considered.

A set-up for the recording of holograms of transparent phase objects consists of a coherent light source, the transparent medium under investigation and optical components, see figure 2.21.

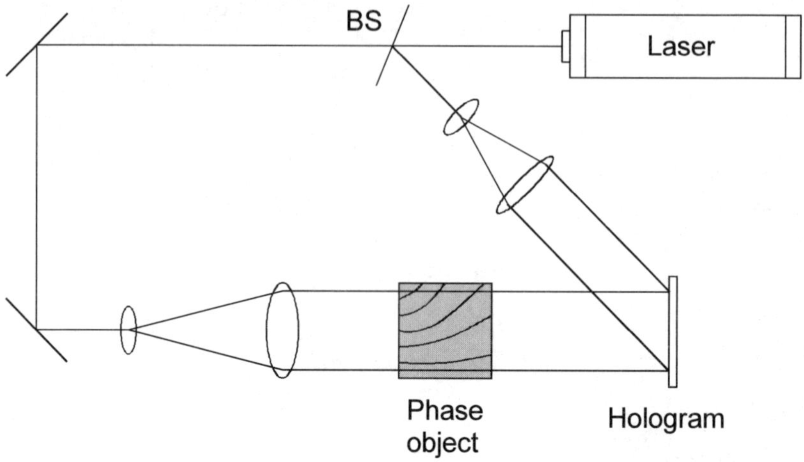


**Fig. 2.21.** Recording set-up for transparent phase objects

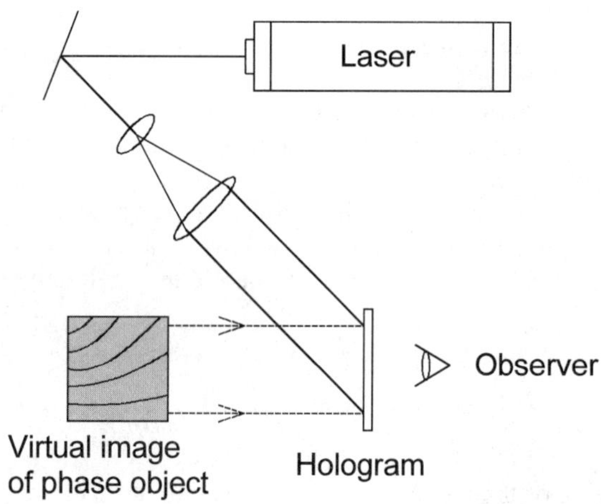


**Fig. 2.22.** Reconstruction of phase objects

The laser beam is split into two partial waves. One partial wave is expanded by a telescopic lens system and illuminates the medium, which is located e. g. in a test cell with transparent walls. The transmitted part, the object wave, interferes with the reference wave at the surface of the hologram plate. After processing the object wave is reconstructed by illuminating the hologram with the reference wave again, figure 2.22. Holographic Interferometry can be done either by the double exposure method or by the real-time method.

A holographic interferogram of a pure transparent object without any scattering parts consists of clear fringes, not disturbed by speckle noise. These fringes are not localized in space, because there are no object contours visible. Yet, for some applications localized fringes are desired. In that case a diffusing screen has to be placed in front of or behind the object volume.

### 2.7.5 Phase Shifting HI

As already mentioned in chapter 2.7.1 it is not possible to calculate $\Delta\varphi$ unambiguously from the measured intensity, because the parameters $A(x,y)$ and $B(x,y)$ in Eq. (2.77) are not known and the sign is not determined.

Phase shifting Holographic Interferometry is a method to determine the interference phase by recording additional information [9, 20, 58, 60]. The principle is to record three or more interference patterns with mutual phase shifts. For the case of three recordings, the interference patterns are described by:

$$\begin{aligned} I_1(x,y) &= A(x,y) + B(x,y)\cos(\Delta\varphi) \\ I_2(x,y) &= A(x,y) + B(x,y)\cos(\Delta\varphi + \alpha) \\ I_3(x,y) &= A(x,y) + B(x,y)\cos(\Delta\varphi + 2\alpha) \end{aligned} \tag{2.98}$$

The equation system can be solved unambiguously for $\Delta\varphi$ if the phase angle $\alpha$ is known (e.g. 120°).

The phase shift can be realized in practice e. g. by a mirror mounted on a piezoelectric translator. The mirror is placed either in the object- or in the reference beam. If appropriate voltages are applied to the piezo during the hologram reconstruction, well defined path changes in the range of fractions of a wavelength can be introduced. These path changes correspond to phase differences between object- and reference wave.

Instead of using the minimum number of three reconstructions with two mutual phase shifts, Eq. (2.98), it is also possible to generate four reconstructions with three mutual phase shifts:

$$\begin{aligned} I_1(x,y) &= A(x,y) + B(x,y)\cos(\Delta\varphi) \\ I_2(x,y) &= A(x,y) + B(x,y)\cos(\Delta\varphi + \alpha) \\ I_3(x,y) &= A(x,y) + B(x,y)\cos(\Delta\varphi + 2\alpha) \\ I_4(x,y) &= A(x,y) + B(x,y)\cos(\Delta\varphi + 3\alpha) \end{aligned} \tag{2.99}$$

In that case the equation system can be solved without knowledge of the phase shift angle $\alpha$, as long as it is constant. The solution for $\Delta\varphi$ is [71]:

$$\Delta\varphi = \arctan\frac{\sqrt{I_1 + I_2 - I_3 - I_4} \cdot \sqrt{3I_2 - 3I_3 - I_1 + I_4}}{I_2 + I_3 - I_1 - I_4} \tag{2.100}$$

Various HI phase shifting methods have been published, which differ in the number of recordings (at least 3), the value of $\alpha$, the way to generate the phase shift (stepwise or continuously) or other details. These methods will not be discussed here in detail. The principle has been described briefly in order to prepare for a comparison of phase determination in conventional HI using photographic plates with the way to obtain phase information in Digital Holographic Interferometry (chapter 4). Finally it is remarked that phase shifting HI is not the only way to determine the phase from a fringe pattern, but it is the most applied method. Other phase evaluating techniques are the Fourier Transform Method, skeletonizing or the heterodyne technique.

### 2.7.6 Phase Unwrapping

Even after having determined the interference phase by a method such as HI phase shifting a problem remains: The cosine function is periodic, i. e. the interference phase distribution is indefinite to an additive integer of $2\pi$:

$$\cos(\Delta\varphi) = \cos(\Delta\varphi + 2\pi n) \qquad n \in Z \tag{2.101}$$

Interference phase maps calculated with the arctan function or other inverse trigonometric functions therefore contain $2\pi$ jumps at those positions, where an extreme value of $\Delta\varphi$ (either $-\pi$ or $\pi$) is reached. The interference phase along a line of such a phase image looks like a saw tooth function, figure 2.23 (a). The correction of these modulo $2\pi$ jumps in order to generate a continuously phase distribution is called *demodulation*, *continuation* or *phase unwrapping*.

Several unwrapping algorithm have been developed in the last years. In the following the so called path-dependent unwrapping algorithm is described. At first a one-dimensional interference phase distribution is considered. The difference between the phase values of adjacent pixels $\Delta\varphi(n+1) - \Delta\varphi(n)$ is calculated. If this difference is less than $-\pi$ all phase values from the n+1 pixel onwards are increased by $2\pi$. If this difference is greater than $+\pi$, $2\pi$ is subtracted from all phase values, starting at number n+1. If none of above mentioned conditions is valid the phase value remains unchanged. The practical implementation of this procedure is done by calculating first a step function, which cumulates the $2\pi$ jumps for all pixels, figure 2.23 (b). The continuous phase distribution is then calculated by adding this step function to the unwrapped phase distribution, figure 2.23 (c). Almost every pixel can be used as a starting point for this unwrapping procedure, not necessarily the pixel at the start of the line. If a central pixel is chosen as starting point the procedure has to be carried out in both directions from that point.

This one-dimensional unwrapping scheme can be transferred to two dimensions. One possibility is to unwrap first one row of the two dimensional phase map with the algorithm described above. The pixels of this unwrapped row act then as a starting points for column demodulation.

One disadvantage of the simple unwrapping procedure described here is that difficulties occur if masked regions are in the phase image. These masked areas might be caused e. g. by holes in the object surface. To avoid this and other difficulties several other, more sophisticated demodulation algorithm have been developed.

Finally it should be mentioned that the unwrapping procedure is always the same for all methods of metrology, which generate saw-tooth like images. This means the various unwrapping algorithm developed for HI and other methods can be used also for Digital Holographic Interferometry, because this technique also generates modulo $2\pi$-images (see chapter 4).

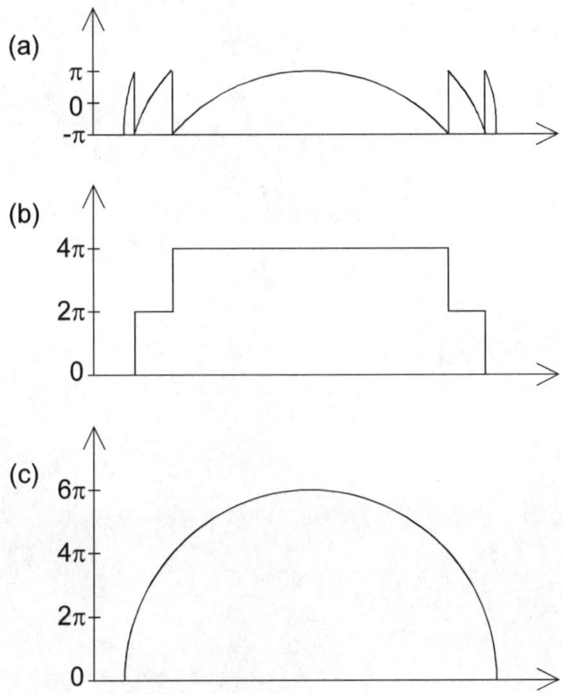


**Fig. 2.23.** Phase unwrapping

(a) Interference phase modulo 2π: $\Delta\varphi_{2\pi}(x)$

(b) Step function: $\Delta\varphi_{jump}(x)$

(c) unwrapped interference phase: $\Delta\varphi_{2\pi}(x) + \Delta\varphi_{jump}(x)$

# 3 Digital Holography

## 3.1 General Principles

The concept of digital hologram recording is illustrated in figure 3.1(a) [131]. A plane reference wave and the wave reflected from the object are interfering at the surface of a Charged Coupled Device (CCD). The resulting hologram is electronically recorded and stored. The object is in general a three dimensional body with diffusely reflecting surface, located at a distance *d* from the CCD.

In optical reconstruction the virtual image appears at the position of the original object and the real image is formed at a distance *d* as well, but in the opposite direction from the CCD, see figure 3.1 (b).

The diffraction of a light wave at an aperture (in this case a hologram) which is mounted perpendicular to the incoming beam is described by the Fresnel-Kirchhoff integral, see Eq. (2.48):

$$\Gamma(\xi',\eta') = \frac{i}{\lambda}\int_{-\infty}^{\infty}\int_{-\infty}^{\infty} h(x,y)E_R(x,y)\frac{\exp\left(-i\frac{2\pi}{\lambda}\rho'\right)}{\rho'}dxdy \tag{3.1}$$

with

$$\rho' = \sqrt{(x-\xi')^2 + (y-\eta')^2 + d^2} \tag{3.2}$$

*h(x,y)* is the hologram function and $\rho'$ is the distance between a point in the hologram plane and a point in the reconstruction plane. The geometrical quantities are explained in figure 3.2. The inclination factor is set to 1, because the angles $\theta$ and $\theta'$ (see chapter 2.4) are approximately 0. This is valid for all numerical reconstruction algorithms in this book.

For a plane reference wave $E_R(x,y)$ is simply given by the real amplitude:

$$E_R = a_R + i0 = a_R \tag{3.3}$$

The diffraction pattern is calculated at a distance *d* behind the CCD plane, which means it reconstructs the complex amplitude in the plane of the real image.

Eq. (3.1) is the basis for numerical hologram reconstruction. Because the reconstructed wave field $\Gamma(\xi',\eta')$ is a complex function, both the intensity as well as the phase can be calculated [128]. This is in contrast to the case of optical hologram reconstruction, in which only the intensity is made visible. This interesting property of Digital Holography is used in Digital Holographic Interferometry, see chapter 4.

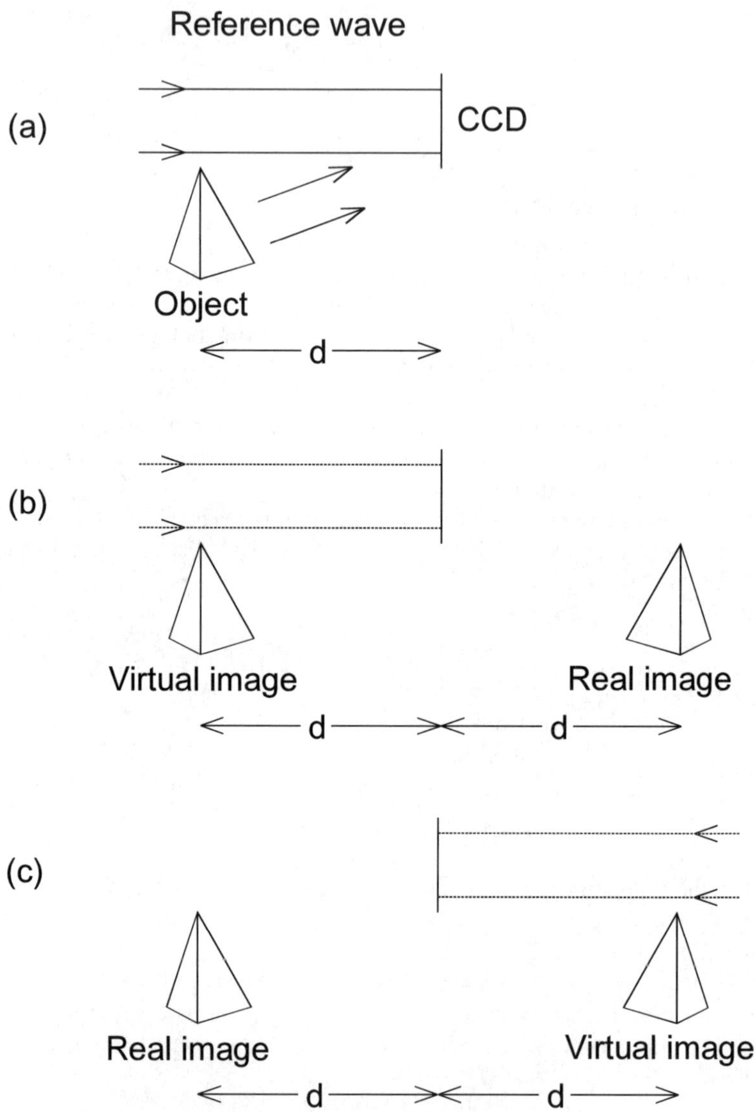


**Fig. 3.1.** Digital Holography
(a) Recording
(b) Reconstruction with reference wave $E_R$
(c) Reconstruction with conjugate reference wave $E_R^*$

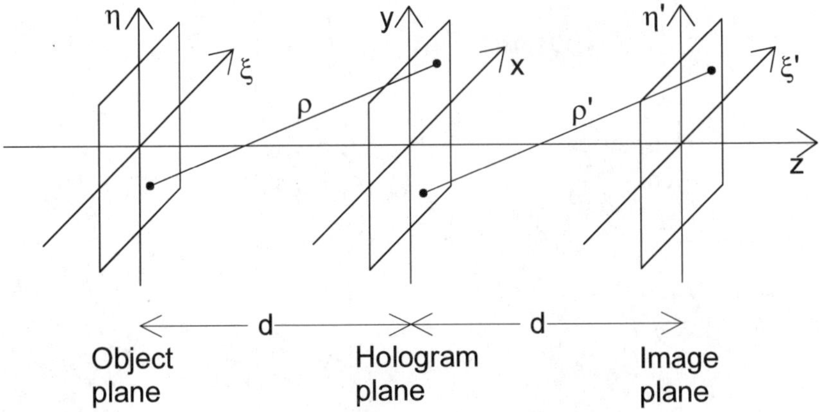


**Fig. 3.2.** Coordinate system for numerical hologram reconstruction

As mentioned in chapter 2.6 the real image could be distorted. According to Eq. (2.63) an undistorted real image can be produced by using the conjugate reference beam for reconstruction. To reconstruct an undistorted real image in Digital Holography it is therefore necessary to insert $E_R^*$ instead of $E_R$ in Eq. (3.1):

$$\Gamma(\xi,\eta)=\frac{i}{\lambda}\int_{-\infty}^{\infty}\int_{-\infty}^{\infty}h(x,y)E_R^*(x,y)\frac{\exp\left(-i\frac{2\pi}{\lambda}\rho\right)}{\rho}dxdy \tag{3.4}$$

with

$$\rho=\sqrt{(x-\xi)^2+(y-\eta)^2+d^2} \tag{3.5}$$

This reconstruction scheme is shown in figure 3.1(c). The real image emerges at that position, where the object was located during recording. It should be mentioned that for the plane reference wave defined in Eq. (3.3) both reconstruction formulas, Eq. (3.1) and (3.4), are equivalent because $E_R = E_R^* \equiv a_R$.

The set-up of figure 3.1 with a plane reference wave impinging perpendicularly onto the CCD is often used in Digital Holography. Other recording geometries are discussed later.

The reconstruction of the virtual image is also possible by introducing the imaging properties of a lens into the numerical reconstruction process [129]. This lens corresponds to the eye lens of an observer watching through an optically reconstructed hologram. In the simplest case this lens is located directly behind the hologram, figure 3.3. The imaging properties of a lens with focal distance $f$ are considered by a complex factor:

$$L(x,y)=\exp\left[i\frac{\pi}{\lambda f}\left(x^2+y^2\right)\right] \tag{3.6}$$

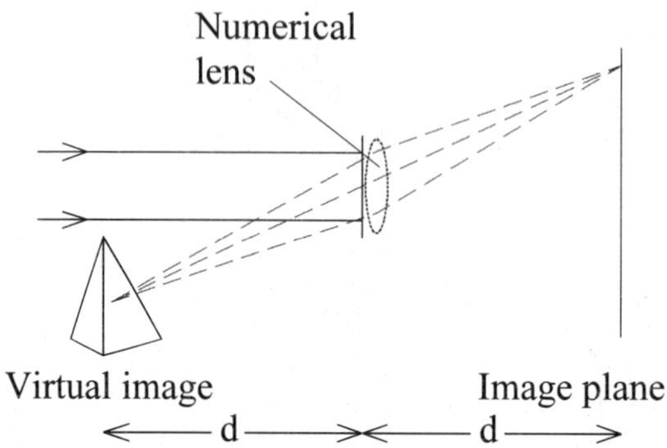


**Fig. 3.3.** Reconstruction of the virtual image

This factor is calculated in Annex B1. For a magnification of 1 a focal distance of $f=d/2$ has to be used.

The lens described by Eq. (3.6) causes phase aberrations, which can be corrected by multiplying the reconstructed wave field by a factor

$$P(\xi',\eta')=\exp\left[i\frac{\pi}{\lambda f}\left(\xi'^2+\eta'^2\right)\right] \tag{3.7}$$

This correction factor is derived in Annex B2. The full formula for reconstruction via a virtual lens with $f=d/2$ is therefore

$$\Gamma(\xi',\eta')=\frac{i}{\lambda}P(\xi',\eta')\int_{-\infty}^{\infty}\int_{-\infty}^{\infty}h(x,y)E_R(x,y)L(x,y)\frac{\exp\left(-i\frac{2\pi}{\lambda}\rho'\right)}{\rho'}dxdy \tag{3.8}$$

## 3.2 Numerical Reconstruction

### 3.2.1 Reconstruction by the Fresnel Approximation

For $x$- and $y$-values as well as for $\xi$- and $\eta$-values, which are small compared to the distance $d$ between the reconstruction plane and the CCD, the expression Eq. (3.5) can be expanded to a Taylor series:

$$\rho = d + \frac{(\xi - x)^2}{2d} + \frac{(\eta - x)^2}{2d} - \frac{1}{8}\frac{\left[(\xi - x)^2 + (\eta - x)^2\right]^2}{d^3} + \ldots \tag{3.9}$$

The fourth term can be neglected, if it is small compared to the wavelength [69]:

$$\frac{1}{8}\frac{\left[(\xi - x)^2 + (\eta - x)^2\right]^2}{d^3} << \lambda \tag{3.10}$$

or

$$d >> \sqrt[3]{\frac{1}{8}\frac{\left[(\xi - x)^2 + (\eta - x)^2\right]^2}{\lambda}} \tag{3.11}$$

Then the distance $\rho$ consists of linear and quadratic terms:

$$\rho = d + \frac{(\xi - x)^2}{2d} + \frac{(\eta - x)^2}{2d} \tag{3.12}$$

With the additional approximation of replacing the denominator in (3.4) by $d$ the following expression results for reconstruction of the real image:

$$\Gamma(\xi,\eta) = \frac{i}{\lambda d}\exp\left(-i\frac{2\pi}{\lambda}d\right) \times \int_{-\infty}^{\infty}\int_{-\infty}^{\infty} E_R^*(x,y)h(x,y)\exp\left[-i\frac{\pi}{\lambda d}\left((\xi - x)^2 + (\eta - y)^2\right)\right]dxdy \tag{3.13}$$

If the multiplications in the argument of the exponential under the integral are carried out one gets

$$\Gamma(\xi,\eta) = \frac{i}{\lambda d}\exp\left(-i\frac{2\pi}{\lambda}d\right)\exp\left[-i\frac{\pi}{\lambda d}\left(\xi^2 + \eta^2\right)\right] \times \int_{-\infty}^{\infty}\int_{-\infty}^{\infty} E_R^*(x,y)h(x,y)\exp\left[-i\frac{\pi}{\lambda d}\left(x^2 + y^2\right)\right]\exp\left[i\frac{2\pi}{\lambda d}(x\xi + y\eta)\right]dxdy \tag{3.14}$$

This equation is named *Fresnel approximation* or *Fresnel transformation* due to its mathematical similarity with the Fourier Transform (see below). It enables reconstruction of the wave field in a plane behind the hologram, in this case in the plane of the real image.

The intensity is calculated by squaring:

$$I(\xi,\eta) = |\Gamma(\xi,\eta)|^2 \tag{3.15}$$

The phase is calculated by

$$\varphi(\xi,\eta) = \arctan\frac{\mathrm{Im}[\Gamma(\xi,\eta)]}{\mathrm{Re}[\Gamma(\xi,\eta)]} \tag{3.16}$$

Re denotes the real part and Im the imaginary part.

The reconstruction formula for the virtual image in the Fresnel approximation is:

$$\begin{aligned}\Gamma(\xi',\eta') &= \frac{i}{\lambda d}\exp\left(-i\frac{2\pi}{\lambda}d\right)\exp\left[-i\frac{\pi}{\lambda d}\left(\xi'^2+\eta'^2\right)\right]P(\xi',\eta') \\ &\times \int_{-\infty}^{\infty}\int_{-\infty}^{\infty} E_R(x,y)L(x,y)h(x,y)\exp\left[-i\frac{\pi}{\lambda d}\left(x^2+y^2\right)\right]\exp\left[i\frac{2\pi}{\lambda d}\left(x\xi'+y\eta'\right)\right]dxdy \\ &= \frac{i}{\lambda d}\exp\left(-i\frac{2\pi}{\lambda}d\right)\exp\left[+i\frac{\pi}{\lambda d}\left(\xi'^2+\eta'^2\right)\right] \\ &\times \int_{-\infty}^{\infty}\int_{-\infty}^{\infty} E_R(x,y)h(x,y)\exp\left[+i\frac{\pi}{\lambda d}\left(x^2+y^2\right)\right]\exp\left[i\frac{2\pi}{\lambda d}\left(x\xi'+y\eta'\right)\right]dxdy\end{aligned} \tag{3.17}$$

For digitisation of the Fresnel transform Eq. (3.14) following substitutions are introduced [176]:

$$\nu = \frac{\xi}{\lambda d}; \qquad \mu = \frac{\eta}{\lambda d} \tag{3.18}$$

Herewith (3.14) becomes

$$\begin{aligned}\Gamma(\nu,\mu) &= \frac{i}{\lambda d}\exp\left(-i\frac{2\pi}{\lambda}d\right)\exp\left[-i\pi\lambda d\left(\nu^2+\mu^2\right)\right] \\ &\times \int_{-\infty}^{\infty}\int_{-\infty}^{\infty} E_R^*(x,y)h(x,y)\exp\left[-i\frac{\pi}{\lambda d}\left(x^2+y^2\right)\right]\exp\left[i2\pi\left(x\nu+y\mu\right)\right]dxdy\end{aligned} \tag{3.19}$$

A comparison of Eq. (3.19) with the definition of the two-dimensional Fourier transform (see Annex A) shows that the Fresnel approximation up to a spherical phase factor is the inverse Fourier transformation of the function $E_R^*(x,y)h(x,y)\exp\left[-i\pi/\lambda d\left(x^2+y^2\right)\right]$:

$$\begin{aligned}\Gamma(\nu,\mu) &= \frac{i}{\lambda d}\exp\left(-i\frac{2\pi}{\lambda}d\right)\exp\left[-i\pi\lambda d\left(\nu^2+\mu^2\right)\right] \\ &\times \Im^{-1}\left\{E_R^*(x,y)h(x,y)\exp\left[-i\frac{\pi}{\lambda d}\left(x^2+y^2\right)\right]\right\}\end{aligned} \tag{3.20}$$

The function $\Gamma$ can be digitised if the hologram function $h(x,y)$ is sampled on a rectangular raster of N×N points, with steps $\Delta x$ and $\Delta y$ along the coordinates. $\Delta x$ and $\Delta y$ are the distances between neighbouring pixels on the CCD in horizontal and vertical direction. With these discrete values the integrals of (3.19) are converted to finite sums:

$$\Gamma(m,n)=\frac{i}{\lambda d}\exp\left(-i\frac{2\pi}{\lambda}d\right)\exp\left[-i\pi\lambda d\left(m^2\Delta\nu^2+n^2\Delta\mu^2\right)\right] \tag{3.21}$$
$$\times\sum_{k=0}^{N-1}\sum_{l=0}^{N-1}E_R^*(k,l)h(k,l)\exp\left[-i\frac{\pi}{\lambda d}\left(k^2\Delta x^2+l^2\Delta y^2\right)\right]\exp\left[i2\pi\left(k\Delta x m\Delta\nu+l\Delta y n\Delta\mu\right)\right]$$

for $m=0,1,\ldots,N-1$; $n=0,1,\ldots,N-1$

According to the theory of Fourier transform among $\Delta x$, $\Delta y$ and $\Delta\nu$, $\Delta\mu$ the following relation exist, see Annex A:

$$\Delta\nu=\frac{1}{N\Delta x};\qquad \Delta\mu=\frac{1}{N\Delta y} \tag{3.22}$$

After re-substitution:

$$\Delta\xi=\frac{\lambda d}{N\Delta x};\qquad \Delta\eta=\frac{\lambda d}{N\Delta y} \tag{3.23}$$

Using these equations (3.21) converts to

$$\Gamma(m,n)=\frac{i}{\lambda d}\exp\left(-i\frac{2\pi}{\lambda}d\right)\exp\left[-i\pi\lambda d\left(\frac{m^2}{N^2\Delta x^2}+\frac{n^2}{N^2\Delta y^2}\right)\right] \tag{3.24}$$
$$\times\sum_{k=0}^{N-1}\sum_{l=0}^{N-1}E_R^*(k,l)h(k,l)\exp\left[-i\frac{\pi}{\lambda d}\left(k^2\Delta x^2+l^2\Delta y^2\right)\right]\exp\left[i2\pi\left(\frac{km}{N}+\frac{ln}{N}\right)\right]$$

This is the discrete Fresnel transform. The matrix $\Gamma$ is calculated by multiplying $E_R^*(k,l)$ with $h(k,l)$ and $\exp\left[-i\pi/(\lambda d)\left(k^2\Delta x^2+l^2\Delta y^2\right)\right]$ and applying an inverse discrete Fourier transform to the product. The calculation is done most effectively using the Fast Fourier Transform (FFT) algorithm. The factor in front of the sum only affects the overall phase and can be neglected, if just the intensity according to Eq. (3.15) is of interest. This is also the case if phase differences between holograms recorded with the same wavelength have to be calculated ($\Delta\varphi=\varphi_1+const.-(\varphi_2+const.)=\varphi_1-\varphi_2$).

The corresponding discrete formula for reconstruction via a virtual lens with $f=d/2$ (Eq. 3.17) is

$$\Gamma(m,n)=\frac{i}{\lambda d}\exp\left(-i\frac{2\pi}{\lambda}d\right)\exp\left[+i\pi\lambda d\left(\frac{m^2}{N^2\Delta x^2}+\frac{n^2}{N^2\Delta y^2}\right)\right]$$
$$\times\sum_{k=0}^{N-1}\sum_{l=0}^{N-1}E_R(k,l)h(k,l)\exp\left[+i\frac{\pi}{\lambda d}\left(k^2\Delta x^2+l^2\Delta y^2\right)\right]\exp\left[i2\pi\left(\frac{km}{N}+\frac{ln}{N}\right)\right] \tag{3.25}$$

A typical digital hologram is shown in figure 3.4. It was recorded with the geometry of figure 3.1. The object is placed *d=1.054m* apart from the CCD-array of 1024×1024 pixels with pixel distances $\Delta x=\Delta y=6.8\mu m$. The wavelength is 632.8nm. The numerical reconstruction according to Eq. (3.14) resp. (3.24) is demonstrated in figure 3.5. A real image of the dice used as object is noticeable. The bright square in the centre is the undiffracted reconstruction wave (zero order) and corresponds to the first term of the right side of Eq. (2.63). Because of the off-axis geometry the image is spatially separated from the zero order term. The other (virtual) image is out of focus in this reconstruction.

An interesting property of holography is that every part of a hologram contains the information about the entire object. This is illustrated by the holograms of figures 3.6 and 3.8, where black masks cover nearly half of the hologram areas. Nevertheless the entire cube is visible in the reconstructions without obstructions, figures 3.7 and 3.9. The masks are visible as shadows in the zero order terms. The reduction of the effective pixel number leads to a reduction of the resolution in the reconstructed images. This corresponds to the increase of the speckle size due to aperture reduction in optical hologram reconstruction.

Regarding Eq. (3.23) the pixel distances in the reconstructed image $\Delta\xi$ and $\Delta\eta$ are depending on the reconstruction distance *d* chosen for numerical reconstruction. This is because Eq. (3.23) corresponds to the diffraction limited resolution of optical systems: The hologram is the aperture of the optical system with side length $N\Delta x$. According to the theory of diffraction a diffraction pattern develops in a distance *d* behind the hologram. $\Delta\xi=\lambda d/N\Delta x$ is therefore the half diameter of the airy disk or speckle diameter in the plane of the reconstructed image, which limits the resolution. This can be regarded as "natural scaling" algorithm, setting the resolution of the image reconstructed by a discrete Fresnel transform always to the physical limit.

**Fig. 3.4.** Digital hologram

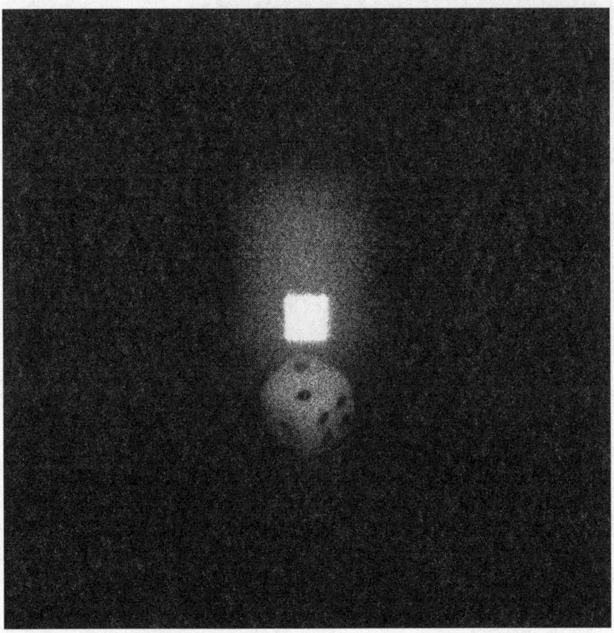

**Fig. 3.5.** Numerical reconstruction

**Fig. 3.6.** Masked digital hologram

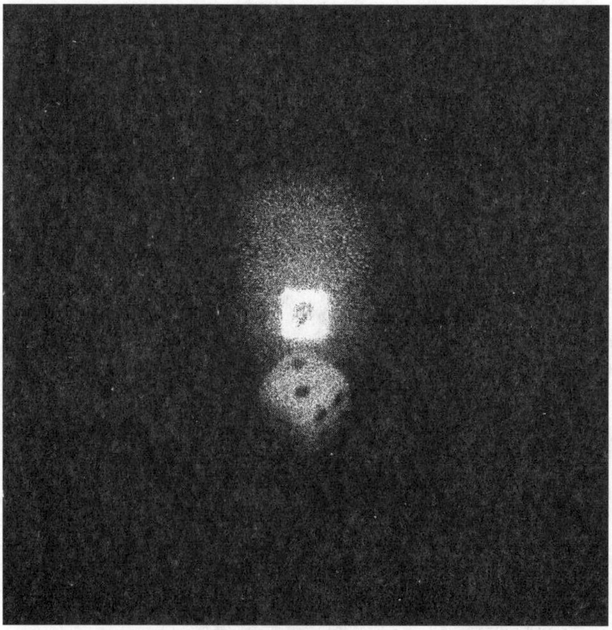

**Fig. 3.7.** Reconstruction

**Fig. 3.8.** Masked digital hologram

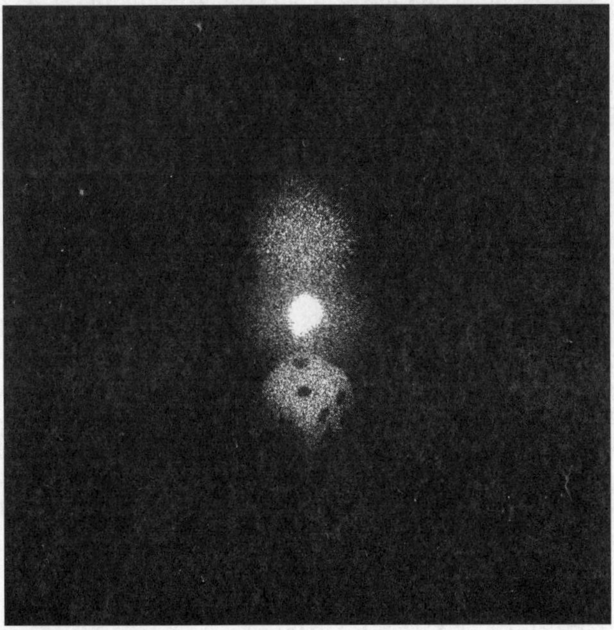

**Fig. 3.9.** Reconstruction

### 3.2.2 Reconstruction by the Convolution Approach

The numerical processing of the Fresnel-Kirchhoff integral Eq. (3.1) resp. (3.4) without approximation is time consuming. For numerical processing a different but equivalent formulation is much more suitable. This formulation makes use of the convolution theorem and is named "convolution approach" within the scope of this article. Demetrakopoulos and Mittra applied this way of processing for numerical reconstruction of sub optical holograms for the first time [25]. Later this approach was applied to optical holography by Kreis [74].

The reconstruction formula Eq. (3.4) can be interpreted as a superposition integral:

$$\Gamma(\xi,\eta)=\int_{-\infty}^{\infty}\int_{-\infty}^{\infty}h(x,y)E_R^*(x,y)g(\xi,\eta,x,y)dxdy \tag{3.26}$$

where the impulse response $g(x,y,\xi,\eta)$ is given by

$$g(\xi,\eta,x,y)=\frac{i}{\lambda}\frac{\exp\left[-i\frac{2\pi}{\lambda}\sqrt{d^2+(x-\xi)^2+(y-\eta)^2}\right]}{\sqrt{d^2+(x-\xi)^2+(y-\eta)^2}} \tag{3.27}$$

According to Eq. (3.26) the linear system characterized by $g(\xi,\eta,x,y)=g(\xi-x,\eta-y)$ is space-invariant. The superposition integral can be regarded therefore as a convolution and the convolution theorem (Annex A) can be applied. According to this theorem the Fourier transform of the convolution of $h\cdot E_R^*$ with $g$ is the product of the individual transforms $\Im\{hE_R^*\}$ and $\Im\{g\}$. So $\Gamma(\xi,\eta)$ can be calculated by Fourier transforming $h\cdot E_R^*$ first. This is followed by multiplying with the Fourier transform of *g*, and taking an inverse Fourier transform of this product. In all three Fourier transforms are necessary for the whole process. The individual Fourier transforms are effectively carried out using the FFT-algorithm.

The numerical impulse response function is

$$g(k,l)=\frac{i}{\lambda}\frac{\exp\left[-i\frac{2\pi}{\lambda}\sqrt{d^2+\left(k-\frac{N}{2}\right)^2\Delta x^2+\left(l-\frac{N}{2}\right)^2\Delta y^2}\right]}{\sqrt{d^2+\left(k-\frac{N}{2}\right)^2\Delta x^2+\left(l-\frac{N}{2}\right)^2\Delta y^2}} \tag{3.28}$$

The shift of the coordinates by *N/2* is on symmetry reasons.

In short notation the reconstruction into the real image plane is

$$\Gamma(\xi,\eta)=\Im^{-1}\left\{\Im\left(h\cdot E_R^*\right)\cdot\Im(g)\right\} \tag{3.29}$$

The Fourier transform of *g(k,l)* can be calculated and expressed analytically:

$$G(n,m)=\exp\left\{-i\frac{2\pi d}{\lambda}\sqrt{1-\frac{\lambda^2\left(n+\frac{N^2\Delta x^2}{2d\lambda}\right)^2}{N^2\Delta x^2}-\frac{\lambda^2\left(m+\frac{N^2\Delta y^2}{2d\lambda}\right)^2}{N^2\Delta y^2}}\right\} \tag{3.30}$$

This saves one Fourier transform for reconstruction:

$$\Gamma(\xi,\eta)=\Im^{-1}\left\{\Im\left(h\cdot E_R^*\right)\cdot G\right\} \tag{3.31}$$

For the reconstruction of the virtual image a lens transmission factor $L(x,y)$ according to Eq. (3.6) and a correction factor $P(\xi',\eta')$ according to Eq. (3.7) have to be considered:

$$\Gamma(\xi',\eta')=P(\xi',\eta')\Im^{-1}\left\{\Im\left(h\cdot E_R\cdot L\right)\cdot G\right\} \tag{3.32}$$

The pixel distances of the images reconstructed by the convolution approach are equal to that of the hologram:

$$\Delta\xi=\Delta x\ ;\qquad \Delta\eta=\Delta y \tag{3.33}$$

The pixel distances of the reconstructed images differ from those of the Fresnel approximation (3.23). At first sight it seems to be possible to achieve a higher resolution with the convolution approach if the pixel distance is small enough. On closer examination one recognizes that the resolution calculated by Eq. (3.33) is only a numerical value. The physical image resolution is determined by diffraction, i. e. Eq. (3.23) describes the resolution limit also for the convolution approach.

As stated in chapter 3.1 the inclination factor in the diffraction integral can be neglected in the numerical reconstruction process. However, it should be mentioned that is also possible to consider the inclination factor in the convolution approach.

The area reconstructed with the impulse response function defined in Eq. (3.28) is symmetrical with respect to the optical axis. The area can be shifted by introducing integer numbers $s_k$, $l_k$:

$$g(k+s_k,l+s_l)=\frac{i}{\lambda}\frac{\exp\left[-i\frac{2\pi}{\lambda}\sqrt{d^2+\left(k-\frac{N}{2}+s_k\right)^2\Delta x^2+\left(l-\frac{N}{2}+s_l\right)^2\Delta y^2}\right]}{\sqrt{d^2+\left(k-\frac{N}{2}+s_k\right)^2\Delta x^2+\left(l-\frac{N}{2}+s_l\right)^2\Delta y^2}} \tag{3.34}$$

The convolution approach contains the possibility to introduce an image magnification in the reconstruction process. This is possible if the reconstruction distance is set to

$$d' = d \cdot m \tag{3.35}$$

where $d$ is the recording distance (also used as reconstruction distance so far) and $m$ stands for the magnification factor. A magnification of $m = 1$ corresponds to $\Delta\xi = \Delta x$ , resp. $\Delta\eta = \Delta y$ . The lens focal distance is given by the lens formula of geometrical optics:

$$f = \left(\frac{1}{d} + \frac{1}{d'}\right)^{-1} \tag{3.36}$$

Now Eq. (3.32) is applied for reconstruction with $d'$ instead of $d$ and

$$L(x,y) = \exp\left[i\frac{\pi}{\lambda f}\left(x^2 + y^2\right)\right] = \exp\left[i\frac{\pi}{\lambda}\left(\frac{1}{d} + \frac{1}{d'}\right)\left(x^2 + y^2\right)\right] \tag{3.37}$$

An example of a reconstruction with the convolution approach is shown in figure 3.10. The hologram of figure 3.4 is reconstructed with a magnification of $m = 1/7$ . The corresponding pixel distance in the reconstructed image is $\Delta\xi = \Delta x / m = 48\mu m$ . The comparison to $\Delta\xi = 96\mu m$ for the Fresnel reconstruction is shown in figure 3.5. That means twice as much pixels are available for the object field. However, it is emphasized again that the physical resolution is the same in figures 3.5 and 3.10.

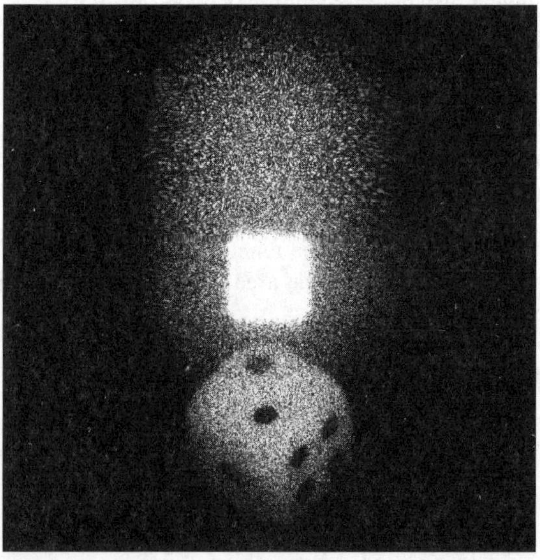

**Fig. 3.10.** Reconstruction with the convolution approach

### 3.2.3 Digital Fourier Holography

The special holographic recording geometry of figure 3.11 is named *lensless Fourier holography*. It also has been realized in Digital Holography [162]. The point source of the spherical reference wave is located in the plane of the object. The reference wave at the CCD plane is therefore described by:

$$E_R = \frac{\exp\left(-i\frac{2\pi}{\lambda}\sqrt{\left(d^2+x^2+y^2\right)}\right)}{\sqrt{\left(d^2+x^2+y^2\right)}} \approx \frac{1}{d}\exp\left(-i\frac{2\pi}{\lambda}d\right)\exp\left(-i\frac{\pi}{\lambda d}\left(x^2+y^2\right)\right) \tag{3.38}$$

The term $\sqrt{d^2+x^2+y^2}$ is the distance between the source point and the point with coordinates *(x,y)* in the CCD plane. The approximation in Eq. (3.38) is the same as used in chapter 3.2.1 to derive the Fresnel transform. Inserting this expression into the reconstruction formula for the virtual image (3.17) leads to following equation:

$$\Gamma(\xi,\eta) = C\exp\left[+i\frac{\pi}{\lambda d}\left(\xi^2+\eta^2\right)\right]\Im^{-1}\{h(x,y)\} \tag{3.39}$$

*C* is a complex constant. A lensless Fourier hologram is therefore reconstructed by a Fourier transform. It is not possible to focus on different areas within the object volume with lensless Fourier holography, because the reconstruction distance *d* does not appear in Eq. (3.39). The spherical phase factor $\exp\left(-i\pi/\lambda d\left(x^2+y^2\right)\right)$ associated with the Fresnel transform is eliminated by the use of a spherical reference wave with the same curvature.

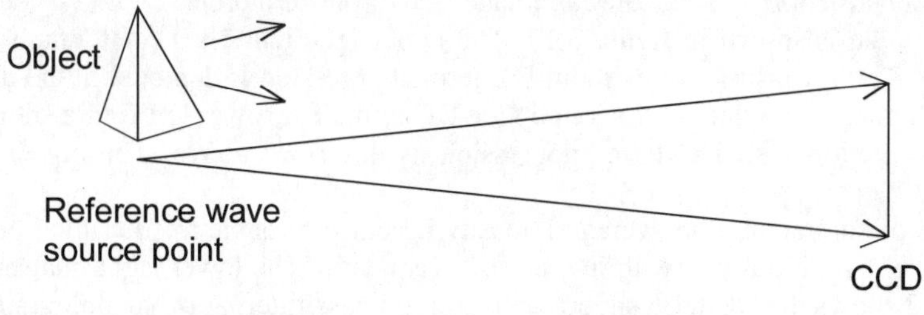


**Fig. 3.11.** Digital lensless Fourier holography

## 3.3 Separation of Virtual Image, Real Image and DC-term

### 3.3.1 Suppression of the DC term

The bright square in the centre of figure 3.5 is the undiffracted reconstruction wave. This zero order or DC term disturbs the image, because it covers all object parts lying behind. Methods have been developed to suppress this term e.g. by Kreis et. al. [75].

To understand the cause of this DC term hologram formation according Eq. (2.60) is considered. The equation is rewritten by inserting the definitions of $E_R$ and $E_O$ and multiplying the terms:

$$
\begin{aligned}
I(x,y) &= \left|E_O(x,y) + E_R(x,y)\right|^2 \\
&= a_R^2 + a_O^2 + 2a_R a_O \cos(\varphi_O - \varphi_R)
\end{aligned}
\tag{3.40}
$$

The first two terms are leading to the DC term in the reconstruction process. The third term is statistically varying between $\pm 2a_R a_O$ from pixel to pixel at the CCD. The average intensity of all pixels of the hologram matrix is

$$
I_m = \frac{1}{N^2}\sum_{k=0}^{N-1}\sum_{l=0}^{N-1} I(k\Delta x, l\Delta y)
\tag{3.41}
$$

The term $a_R^2 + a_O^2$ now can be suppressed by subtracting this average intensity $I_m$ from the hologram:

$$
I'(k\Delta x, l\Delta y) = I(k\Delta x, l\Delta y) - I_m(k\Delta x, l\Delta y)
\tag{3.42}
$$

for $k = 0,\ldots,N-1;\quad l = 0,\ldots,N-1$.

The reconstruction of $I'$ creates an image free of the zero order.

An example is shown in figure 3.12. The upper left figure is a photograph of the object. The reconstruction without DC term suppression is depicted in the upper right figure. The object is covered by the DC term. The lower left figure shows the reconstruction with DC term suppression as described above. The object is clearly visible.

Instead of subtracting the average intensity it is also possible to filter the hologram matrix by a high-pass with low cut-off frequency. The lower right image of figure 3.12 shows the DC term suppression by high pass filtering of the hologram.

The subtraction of the average intensity from the hologram before reconstruction is the basic idea of the DC-term suppression. The same effect can be achieved, if two holograms with stochastically changed speckle structures are subtracted from each other [26]. The reconstruction of this subtraction hologram results in an image without zero order term.

**Fig. 3.12.** Suppression of the DC term (courtesy of S. Seebacher)

Another method to suppress the DC term is to measure the intensities of reference wave $a_R^2$ and object wave $a_O^2$ separately. This can be done e. g. by blocking one of the waves while measuring the intensity of the other wave. Afterwards a DC term free image can be calculated by subtracting the intensities from the hologram before reconstruction. However, this requires higher experimental effort due to the additional measurements.

### 3.3.2 Spatial Separation of Images

With the recording geometry of figure 3.1 the real image and the virtual image are located in one line. In the numerical reconstruction one focuses either on the real or on the virtual image. The other image is usually out of focus due to the long distance between object and CCD. Consequently only one image is visible in the reconstruction, see example in figure 3.5.

Still, there are some cases, in which the reconstruction might be disturbed by the out of focus twin image, e. g. if the recording distance is very short. For this special case it might be useful to record the holograms with a tilted reference, see set-up in figure 3.13. In this geometry both images are angularly separated in the reconstruction. The real image is deflected off the axis at an angle approximately twice as large as the one between reference wave and axis.

The tilted reference wave is described by

$$E_R = \exp\left(-i\frac{2\pi}{\lambda}x\sin\theta\right) \tag{3.43}$$

The disadvantage of this set-up are the much higher spatial frequencies on the CCD in comparison to the set-up of figure 3.1. This is critical with respect to the CCD resolution and limits the usable angle for the object wave, see discussion in chapter 3.4.

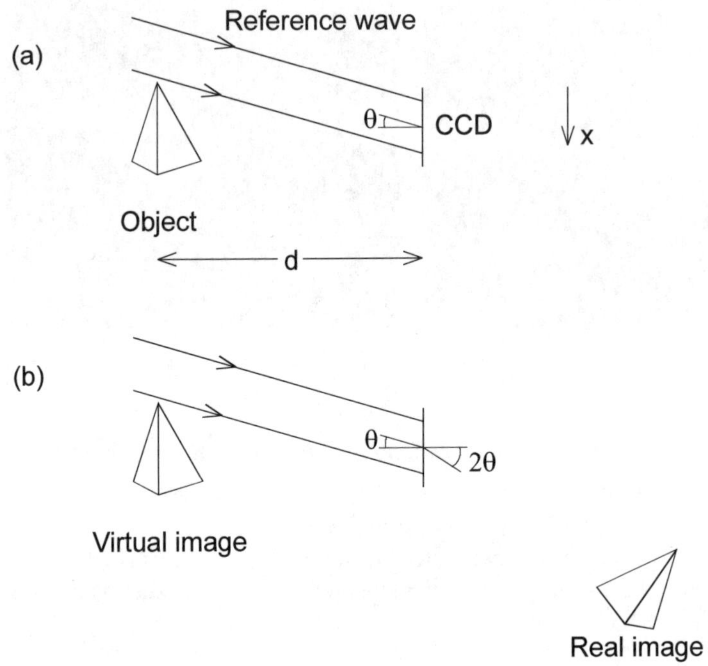


**Fig. 3.13.** Digital Holography with a tilted reference wave
(a) Recording
(b) Reconstruction

### 3.3.3 Phase Shifting Digital Holography

The amplitude and phase of a light wave can be reconstructed from a single hologram by the methods described in the preceding chapters. A completely different approach has been proposed by Skarman [146, 165]. He used a phase shifting algorithm to calculate the *initial* phase and thus the complex amplitude in any plane, e. g. the image plane: With the initial complex amplitude distribution in one plane the wave field in any other plane can be calculated by using the Fresnel-Kirchhoff formulation of diffraction. Later this Phase Shifting Digital Holography was improved and applied to opaque objects by Yamaguchi et al. [54, 171-174, 178, 179].

The principal arrangement for phase shifting Digital Holography is shown in figure 3.14. The object wave and the reference wave are interfering at the surface of a CCD. The reference wave is guided via a mirror mounted on a piezoelectric transducer (PZT). With this PZT the phase of the reference wave can be shifted stepwise. Several (at least three) interferograms with mutual phase shifts are recorded. Afterwards the object phase $\varphi_O$ is calculated from these phase shifted interferograms, the procedure is similar to that of phase shifting in conventional HI (see chapter 2.7.5). The real amplitude $a_O(x,y)$ of the object wave can be measured from the intensity by blocking the reference wave.

As a result the complex amplitude

$$E_O(x,y) = a_O(x,y)\exp(+i\varphi_O(x,y)) \tag{3.44}$$

of the object wave is determined in the recording (x,y) plane.

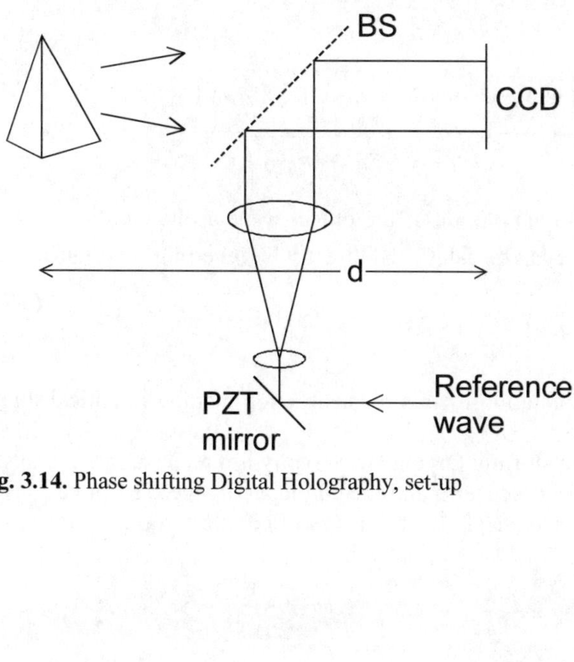


**Fig. 3.14.** Phase shifting Digital Holography, set-up

Now the Fresnel-Kirchhoff integral can be used to calculate the complex amplitude in any other plane. To calculate an image of the object an artificial lens with $f = d/2$ is introduced in the recording plane according to Eq. (3.6). By means of the Fresnel approximation Eq. (3.17) the complex amplitude in the image plane is then calculated by

$$\begin{aligned} E_O(\xi',\eta') &= C\exp\left[+i\frac{\pi}{\lambda d}\left(\xi'^2+\eta'^2\right)\right] \\ &\times \int_{-\infty}^{\infty}\int_{-\infty}^{\infty} E_O(x,y)L(x,y)\exp\left[-i\frac{\pi}{\lambda d}\left(x^2+y^2\right)\right]\exp\left[i\frac{2\pi}{\lambda d}\left(x\xi'+y\eta'\right)\right]dxdy \\ &= C\exp\left[+\frac{i\pi}{\lambda d}\left(\xi'^2+\eta'^2\right)\right] \\ &\times \int_{-\infty}^{\infty}\int_{-\infty}^{\infty} E_O(x,y)\exp\left[+i\frac{\pi}{\lambda d}\left(x^2+y^2\right)\right]\exp\left[i\frac{2\pi}{\lambda d}\left(x\xi'+y\eta'\right)\right]dxdy \end{aligned} \tag{3.45}$$

where again the coordinate system of figure 3.2 is applied. Since the complex amplitude is known in the hologram plane, it is also possible to reconstruct the object by inversion of the recording process [142]. Hologram recording is described by

$$\begin{aligned} E_O(x,y) &= \frac{i}{\lambda}\int_{-\infty}^{\infty}\int_{-\infty}^{\infty} E_O(\xi,\eta)\frac{\exp\left(-i\frac{2\pi}{\lambda}\sqrt{d^2+(\xi-x)^2+(\eta-y)^2}\right)}{\sqrt{d^2+(\xi-x)^2+(\eta-y)^2}}d\xi d\eta \\ &= \Im^{-1}\left\{\Im\left(E_O(\xi,\eta)\right)\cdot\Im\left(g(\xi,\eta,x,y)\right)\right\} \end{aligned} \tag{3.46}$$

with

$$g(\xi,\eta,x,y) = \frac{i}{\lambda}\frac{\exp\left(-i\frac{2\pi}{\lambda}\sqrt{d^2+(\xi-x)^2+(\eta-y)^2}\right)}{\sqrt{d^2+(\xi-x)^2+(\eta-y)^2}} \tag{3.47}$$

$E_O(\xi,\eta)$ is the complex amplitude of the object wave at the emitting surface, see figure 3.2. Therefore it can be calculated directly by inverting Eq. (3.46):

$$E_O(\xi,\eta) = \Im^{-1}\left\{\frac{\Im\left(E_O(x,y)\right)}{\Im\left(g(\xi,\eta,x,y)\right)}\right\} \tag{3.48}$$

The numerical implementation of this reconstruction method is critical due to the division in Eq. (3.48).

The advantage of phase shifting Digital Holography is a reconstructed image of the object without the zero order term and the conjugate image. The price for this achievement is the higher technical effort: Phase shifted interferograms have to be generated, restricting the method to slowly varying phenomena with constant phase during the recording cycle.

Phase shifting Digital Holography is illustrated by a holographic image of a nut, figure 3.15. The example demonstrates the improvement compared to conventional Digital Holography, figure. 3.12.

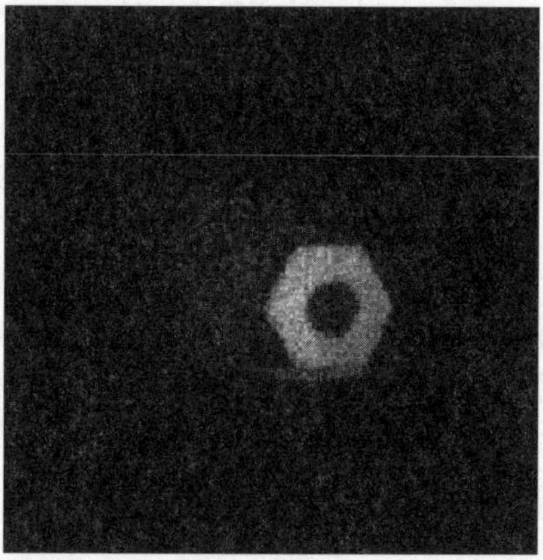

**Fig. 3.15.** Phase shifting Digital Holography, example

## 3.4 Recording of Digital Holograms

### 3.4.1 Charged-Coupled Devices

Charged-Coupled Devices (CCD's) were invented in the sixties of the last century by researchers at Bell Labs. A CCD is an electrical device that is used to create images of objects, store information or transfer electrical charge. The most popular application today is image recording. Therefore the following overview in this book is restricted to this application. CCD's are used as imaging devices in electronic cameras (video and still imaging) and scanners. They are available as line scanning devices, consisting of a single line of light detectors, and as area scanning devices, consisting of a rectangular matrix of light detectors. For Digital Holography only the latter architecture is of interest.

CCD imaging is performed in a three step process [14]:

1. Light exposure
   The incident light separates charges by the internal photo effect. This effect converts light into an electronic charge at the individual detectors called pixels.
2. Charge transfer
   The charge transfer function moves the packets of charge within the semiconductor (silicon) substrate to memory cells.
3. Charge to voltage conversion and output amplification

The capacitor matrix of the memory cells converts the transferred charge to a voltage. An amplifier adapts the voltage to the output requirements.

The three basic architectures of CCD's are called full-frame transfer devices, frame-transfer devices and interline transfer devices.

Interline transfer devices consist of a striped pattern of light sensitive detectors (photodiodes) and of separated non-sensitive or light shielded storage devices, see figure 3.16. The charge packages which are generated in the light sensitive pixels are shifted into the adjacent storage devices by a parallel clock. After that the charges in these storage devices are shifted line by line into the serial register. This serial register transfers the charge packages into a charge to voltage converter with amplifier, which forms the output signal. The major disadvantage of interline transfer CCD's is their complexity, which results from separating the photodetecting and storage (readout) functions.

Frame-transfer CCD's also have different areas for light conversion and for storage. Yet, these elements are not arranged in stripes, but divided into two different blocks, see figure 3.17. The idea is to shift a captured scene from the photosensitive image array very fast to the storage array. The readout from the storage register is performed similar to the readout process of interline transfer devices.

Full-Frame CCD's have the simplest architecture, see figure 3.18. In contrast to interline transfer and frame-transfer devices they don't have a separated storage area. The entire sensor area is light sensitive. The photons of the illuminating light are converted into charge packages. The resulting rows of image information are then shifted in parallel to the serial register that subsequently shifts the row of information to the output as a serial stream of data. The process repeats until all rows are transferred off chip. Since the parallel register is used for both image detection and readout, a mechanical shutter is needed to preserve scene integrity. Yet, full-frame CCD's have highest resolution and the production costs are comparably cheap.

In principle all three types of CCD's are suitable for Digital Holography. Full frame type CCD's have the advantage that the exposure time can be adjusted according to the demands of a specific application. Even exposure times in the range of seconds are possible. However, the mechanical shutter limits the number of holograms, which can be recorded per second (frame rate). In addition the shutter may cause mechanical vibrations to the set-up, which deteriorate the hologram quality. An advantage of interline transfer type CCD's is that such devices are equipped with an electronic shutter, allowing higher frame rates. The best suited camera type depends therefore on the specific holographic application.

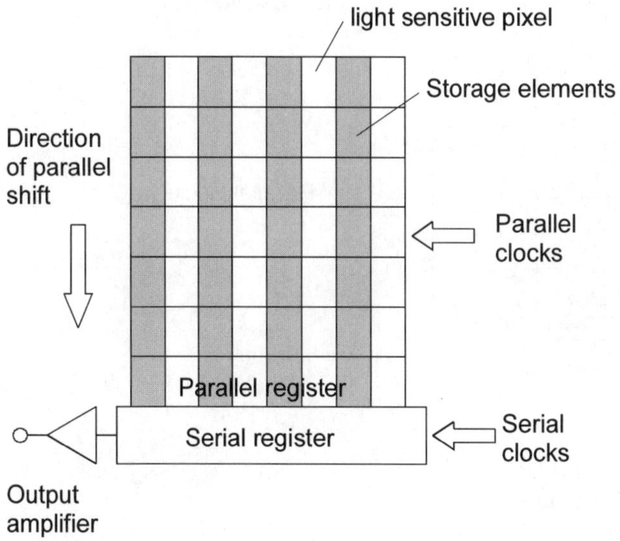


**Fig. 3.16.** Interline-transfer architecture

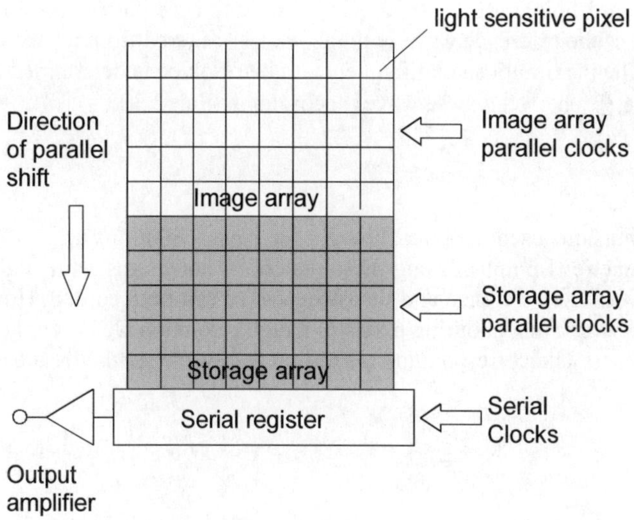


**Fig. 3.17.** Frame-transfer architecture

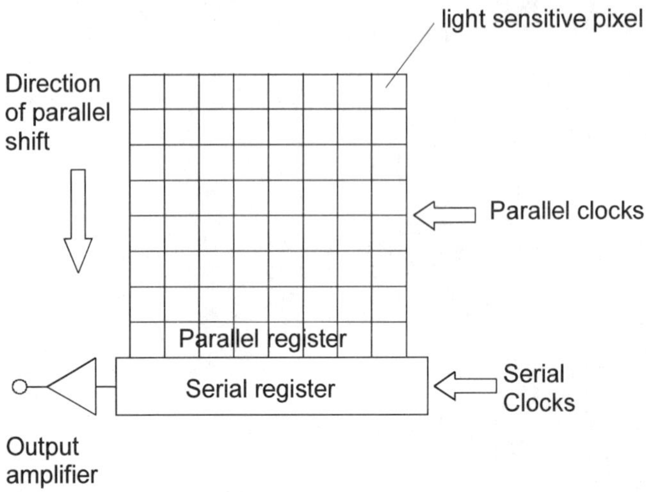


**Fig. 3.18.** Full-frame architecture

### 3.4.2 Spatial Frequency Requirements

A CCD used to record holograms must resolve the interference pattern resulting from superposition of the reference wave with the waves scattered from the different object points. The maximum spatial frequency to be resolved is determined by the maximum angle $\theta_{max}$ between these waves according to Eq. (2.30):

$$f_{\max} = \frac{2}{\lambda}\sin\frac{\theta_{\max}}{2} \tag{3.49}$$

Photographic emulsions used in optical holography have resolutions up to 5000 line pairs per millimetre (Lp/mm). Using these materials, holograms with angles of up to 180° between the reference and the object wave can be recorded. However, the distance between neighbouring pixels of a high resolution CCD is only in the order of $\Delta x \approx 5\mu m$. The corresponding maximum resolvable spatial frequency calculated by

$$f_{\max} = \frac{1}{2\Delta x} \tag{3.50}$$

is therefore in the range of 100 Lp/mm.
Combining Eq. (3.49) and (3.50) leads to

$$\theta_{\max} = 2\arcsin\left(\frac{\lambda}{4\Delta x}\right) \approx \frac{\lambda}{2\Delta x} \tag{3.51}$$

where the approximation is valid for small angles. The distance between neighboring pixels is therefore the quantity, which limits the maximum angle between reference- and object wave.

### 3.4.3 CCD's for Digital Hologram Recording

Main data of selected CCD cameras suitable for Digital Holography are listed in Table 3.1.

**Table 3.1.** CCD cameras suitable for Digital Holography

| Camera | Chip type | Number of pixels | Pixel size [$\mu m^2$] | Frames per second | Dynamic range | $\theta_{max}$ for $\lambda$=633nm |
|---|---|---|---|---|---|---|
| Roper Sci. MegaPlus 1.4i | FT | 1317 x 1035 | 6.8 x 6.8 | 6.9 | 8 bit | 2.7° |
| Roper Sci. MegaPlus 16.8i | FT | 4096 x 4096 | 9 x 9 | 0.47 | 8 bit | 2.0° |
| Roper Sci. MegaPlus ES1.0 | IT | 1008 x 1018 | 9 x 9 | 30 | 8 or 10 bit | 2.0° |
| Roper Sci. MegaPlus 4.0 | IT | 2048 x 2048 | 7.4 x 7.4 | 30 | 8 or 12 bit | 2.45° |
| Hamamatsu C8484-01 | PSI | 1344 x 1024 | 6.45 x 6.45 | 8.3 | 12 bit | 2.81° |
| Duncan DT1100 | PS | 1392 x 1040 | 4.65 x 4.65 | 12 | 8 or 10 bit | 3.9° |

FT: full frame, IT: interline transfer, PSI: progressive scan interline, PS: progressive scan

The sensitivity of CCD cameras is typically in the range of $10^{-4}\,J/m^2$ to $10^{-3}\,J/m^2$, which is better than the sensitivity of photographic emulsions used for classical holography. The spectral range reaches approximately from 400*nm* to 1000*nm*, based on silicon as chip material. Yet, most commercial CCD cameras are equipped with a filter, restricting the sensitivity to the visible range.

In conventional holography with photographic plates the intensity ratio between reference and object wave should be in the range of 5:1 to 10:1 in order to avoid nonlinear effects due to the recording medium. However, from interference theory it is known that the maximum contrast in an interference pattern is achieved if the intensity ratio is 1:1. CCD's have a much better linearity in the exposure curve than photographic emulsions. Consequently, the intensity ratio between reference and object wave should be adjusted to the optimum of 1:1. This can be controlled by covering one half of the expanded reference wave by an aperture. The brightness (measured in grey values) in that half of the CCD being illuminated by reference and object wave together should be twice as high of the brightness in the other half, which is illuminated only by the object wave.

As for classical holography using photographic material, the total light energy impinging on the CCD can be controlled by varying the exposure time. This is usually done with a mechanical or the electronic camera shutter.

CCD-cameras typically have a dynamic range of 8 bit (256 grey values) or higher. This is comparable to the dynamics of photographic materials and fully sufficient for hologram recording. Even objects with brightness variations exceeding the dynamic range of the recording medium by far can be stored and reconstructed, because the object information is coded as interference pattern (hologram).

Numerical reconstruction of digital holograms requires a pixel number, which is a power of 2 (e. g. 1024 x 1024). The pixel numbers of some of the cameras listed in table 3.1 differ from that rule. If the pixel number is e. g. 1317 x 1035 (MegaPlus 1.4i) only 1024 x 1024 pixels are used for reconstruction. In case of a pixel number slightly lower than a power of 2 (e. g. 1008 x 1018 for the MegaPlus ES 1.0) it is advisable to add artificial pixels with grey value zero (black) to the recorded hologram until a pixel number of $2^n \times 2^n$ is reached. This *zero padding* does not distort the computation, it only causes a smoothing or interpolation of the reconstructed image.

### 3.4.4 Recording Set-ups

In this chapter typical set-ups used in Digital Holography are discussed and analysed with respect to their spatial frequency limitations. In figure 3.19(a) a plane reference wave according to Eq. (3.3) is used, which propagates perpendicularly to the CCD. The object is located unsymmetrical with respect to the centre line. This set-up is very simple, but the space occupied by the object is not used effectively. The maximum angle between a ray emitted from the edge of a quadratic object with side length $L$ to the opposite edge of the CCD with side length $N\Delta x$ is (distance $x$ explained in figure 3.19):

$$\theta_{\max} \approx \frac{x}{d_{\min}} = \frac{\sqrt{\frac{5}{4}}\left(L + N\Delta x\right)}{d_{\min}} \tag{3.52}$$

The corresponding minimum object distance $d_{min}$ is calculated by equating this expression with the approximation for $\theta_{\max}$ in Eq. (3.51):

$$d_{\min} = \sqrt{\frac{5}{4}}\frac{2\Delta x}{\lambda}\left(L + N\Delta x\right) = \sqrt{5}\frac{\Delta x}{\lambda}\left(L + N\Delta x\right) \tag{3.53}$$

In figure 3.19(b) the plane reference wave is coupled into the set-up via a beam splitter. This allows positioning the object symmetrically, i.e. objects with larger dimensions can be recorded at a given distance d. The minimum object distance is:

$$d_{\min} \approx \frac{x}{\theta_{\max}} = \sqrt{2}\frac{\Delta x}{\lambda}(L + N\Delta x) \tag{3.54}$$

However, the DC term is in the centre of the reconstructed image and has to be suppressed by the procedures described in chapter 3.3.1.

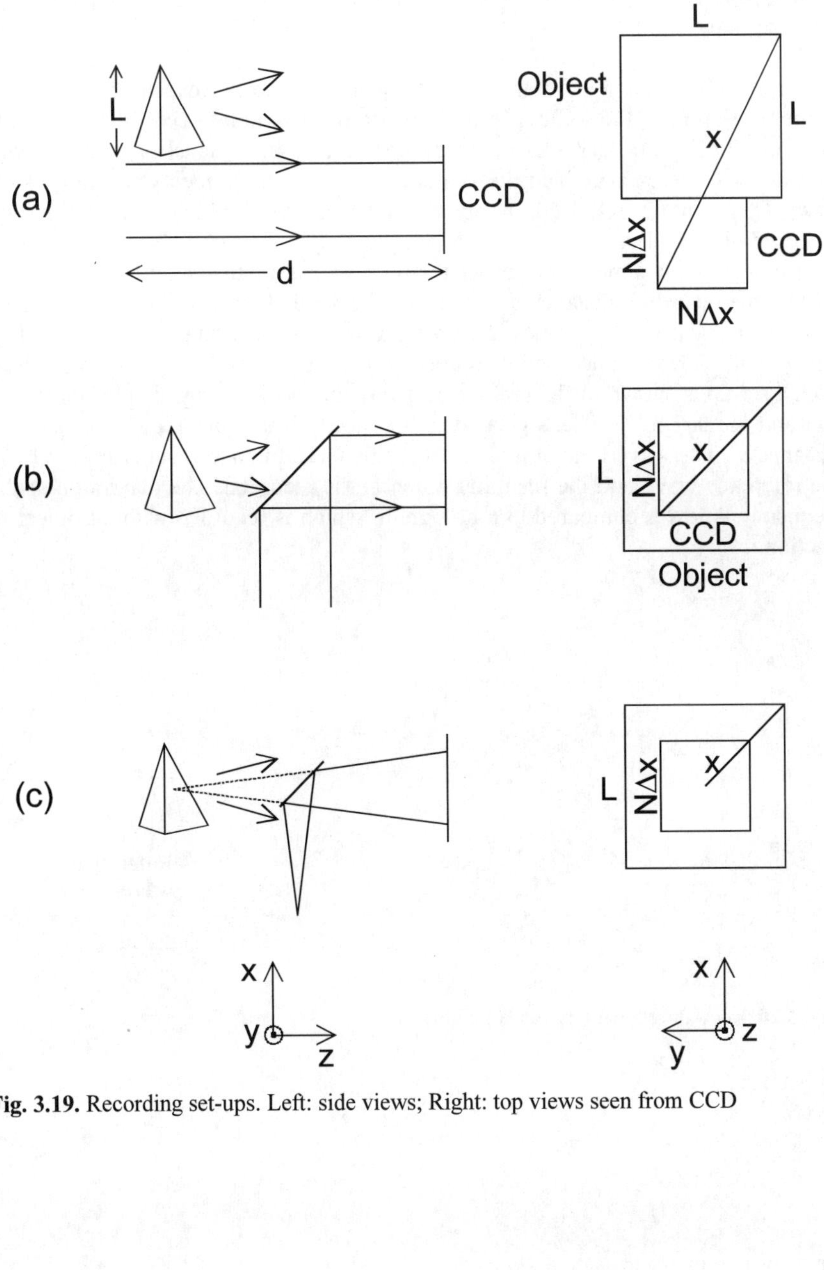


**Fig. 3.19.** Recording set-ups. Left: side views; Right: top views seen from CCD

Figure 3.19(c) shows a set-up for lensless Fourier holography. The spherical reference wave is coupled into the set-up via a beam splitter in order to have the source point in the object plane. The minimum object distance is:

$$d_{\min} = \sqrt{2}\frac{\Delta x}{\lambda}L \tag{3.55}$$

With the lensless Fourier set-up the shortest object distance can be chosen. On the other hand numerical refocusing is not possible with this geometry (see chapter 3.2.3).

For all set-ups the maximum spatial frequency has to be adapted very carefully to the resolution of the CCD. If too high spatial frequencies occur, the contrast of the entire hologram decreases or, in the extreme case, it vanishes totally. Therefore it has to be ensured that the spatial frequency requirements are kept in any case. This can be achieved in practice by apertures, which restrict the light propagation laterally.

If objects with dimensions larger than a few centimetres shall be recorded by CCD's, the recording distance *d* increases up to several meters. This is not feasible in practice. That is why set-ups have been developed to reduce the object angle to values with a resolvable spatial frequency spectrum [118, 139]. An example of such a set-up is shown in figure 3.20. A divergent lens is arranged between the object and the target. This lens generates a reduced virtual image of the object in a distance *d'*. The wave field emerging from this virtual image is superimposed with the reference wave and the resulting hologram is recorded. The maximum spatial frequency is lower compared to a hologram, which is recorded without object reduction.

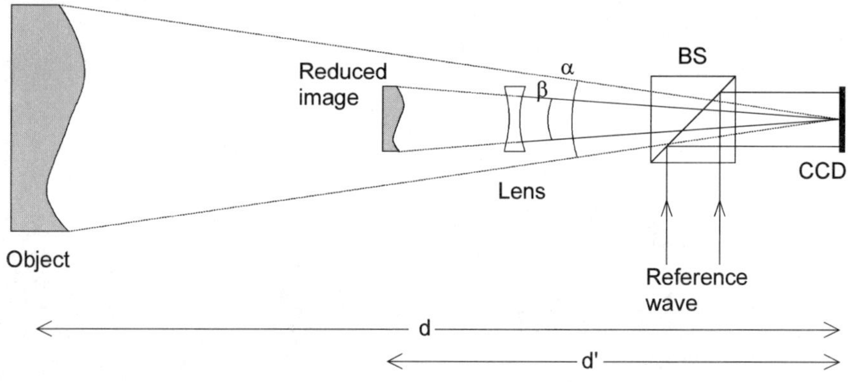


**Fig. 3.20.** Recording geometry for large objects

### 3.4.5 Stability Requirements

A stable optical set-up is necessary for digital as well as for conventional holography. Any change in the optical path difference between the interfering beams will result in a movement of the fringes and reduced contrast in the hologram. In practice the path variations should not exceed 1/4, better 1/10 of a wavelength during hologram exposure. It is therefore recommended to mount the optical set-up on a vibration isolated table, or to use a pulsed laser as light source. In contrast to conventional holography disturbances due to vibrations are visible in Digital Holography even in the recording process: The hologram visible on the monitor of the recording system appears flat and has a low modulation. This is an easy way to monitor the stability of the set-up against vibrations.

# 4 Digital Holographic Interferometry (DHI)

## 4.1 General Principles

A conventional holographic interferogram is generated by superposition of two waves, which are scattered from an object in different states. The interferogram carries the information about the phase change between the waves in form of dark and bright fringes. However, as described in chapter 2.7, the interference phase cannot be extracted unambiguously from a single interferogram. The interference phase is usually calculated from three or more phase shifted interferograms by phase shifting algorithm. This requires additional experimental effort.

Digital Holography allows a completely different way of processing [128]. In each state of the object one digital hologram is recorded. Instead of superimposing these holograms as in conventional HI using photographic plates, the digital holograms are reconstructed separately according to the theory of chapter 3. From the resulting complex amplitudes $\Gamma_1(\xi,\eta)$ and $\Gamma_2(\xi,\eta)$ the phases are calculated:

$$\varphi_1(\xi,\eta) = \arctan\frac{\operatorname{Im}\Gamma_1(\xi,\eta)}{\operatorname{Re}\Gamma_1(\xi,\eta)} \tag{4.1}$$

$$\varphi_2(\xi,\eta) = \arctan\frac{\operatorname{Im}\Gamma_2(\xi,\eta)}{\operatorname{Re}\Gamma_2(\xi,\eta)} \tag{4.2}$$

The index 1 denotes the first state, the index 2 is for the second state. In Eq. (4.1) and (4.2) the phase takes values between $-\pi$ and $\pi$, the principal values of the arctan function. The interference phase is now calculated directly by subtraction:

$$\Delta\varphi = \begin{cases} \varphi_1 - \varphi_2 & if\ \varphi_1 \geq \varphi_2 \\ \varphi_1 - \varphi_2 + 2\pi & if\ \varphi_1 < \varphi_2 \end{cases} \tag{4.3}$$

This equation permits the calculation of the interference phase modulo $2\pi$ directly from the digital holograms. The generation and evaluation of an interferogram is not necessary.

The scheme of Digital HI is shown in figure 4.1. The upper left and upper right figures present two digital holograms, recorded in different states. Between the two recordings the knight has been tilted by a small amount. Each hologram is reconstructed separately by a numerical Fresnel transform. The reconstructed phases according to Eq. (4.1) and (4.2) are depicted in the two figures of the middle row.

The phases vary randomly due to the surface roughness of the object. Subtraction of the phases according to Eq. (4.3) results in the interference phase, lower left figure.

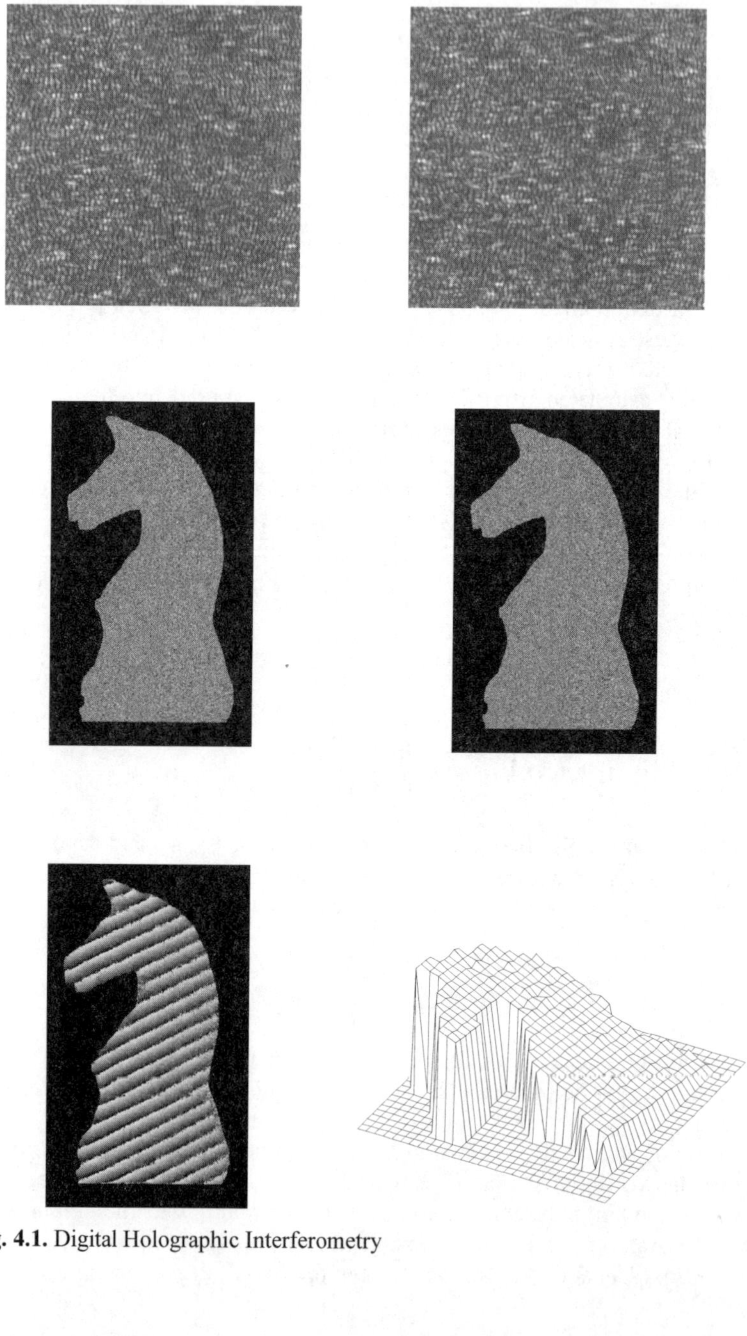

**Fig. 4.1.** Digital Holographic Interferometry

The interference phase is indefinite to an additive multiple of $2\pi$, i. e. it is wrapped modulo $2\pi$. The information about the additive constant is already lost in the holographic interferometric process. This is not a special property of Digital HI, but also the case for all interferometric methods using the wavelength as a length unit. To convert the interference phase modulo $2\pi$ into a continuous phase distribution, one can apply the standard phase unwrapping algorithm developed for conventional interferometry, HI or ESPI. In this example a simple path dependent unwrapping algorithm, as described in chapter 2.7.6, has been applied. The unwrapped phase image is shown in the lower right picture of figure 4.1. The sensitivity vector used for hologram recording is nearly constant and perpendicular over the whole surface. The grey values of the unwrapped phase map can be converted therefore directly into displacement values via Eq. (2.86), i. e. the plot in the lower right picture of figure 4.1 is the object displacement.

## 4.2 Deformation Measurement

### 4.2.1 Quantitative Displacement Measurement

As discussed in chapter 4.1 the way to obtain the interference phase in DHI is totally different from conventional HI using photographic recording media and optical reconstruction. On the other hand the derivation of the relation between displacement vector $\vec{d}$, geometrical boundary conditions described by the sensitivity vector $\vec{S}$ and interference phase $\Delta\varphi$ is also valid for DHI. That means the deformation is calculated by Eq. (2.84), which is repeated here:

$$\Delta\varphi(x,y) = \frac{2\pi}{\lambda}\vec{d}(x,y,z)\left(\vec{b}-\vec{s}\right) = \vec{d}(x,y,z)\vec{S} \tag{4.4}$$

As an example for quantitative displacement measurement the deformation of a plate due to impact loading is discussed [134, 139]. The plate is made of fibre reinforced plastic (FRP), which is used more and more in aircraft industry. The deformation behaviour of FRP under impact load differs from that of static loading, so impact experiments are necessary. DHI is well suited to measurement of such transient deformations, because only one single recording is necessary in each deformation state.

The holographic set-up is shown in figure 4.2. The dimensions of the plate are 12cm × 18cm. The recording distance would be too long for direct recording. The spatial frequency spectrum is therefore reduced by a lens, as explained in chapter 3.4.4 (set-up in figure 3.20).

The plate is clamped at three sides by a stable frame. A pneumatically accelerated steel projectile hits the plate and causes a transient deformation. Two holograms are recorded: The first exposure takes place before the impact, when the plate is in rest. The second hologram is recorded 5 microseconds after the impact.

The holograms are recorded by a pulsed ruby laser with a pulse duration of about 30ns, short enough for hologram recording of dynamic events. The second hologram recording is triggered by a photoelectric barrier, which generates the start signal for the laser after the projectile has crossed. The time interval between the impact and the second laser pulse is adjustable by an electronic delay.

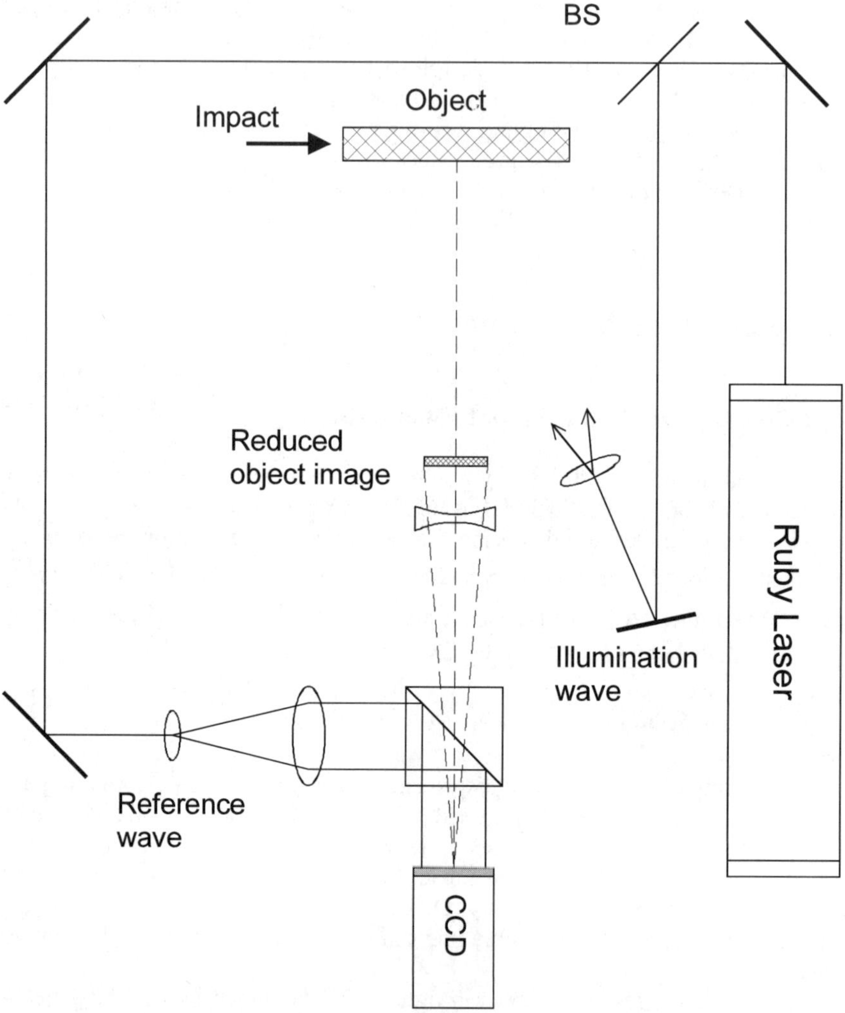


**Fig. 4.2.** Measurement of transient deformations

**Fig. 4.3.** Interference phase modulo $2\pi$

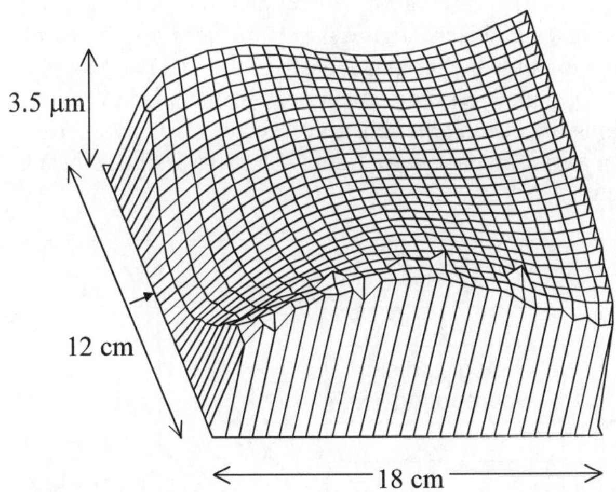


**Fig. 4.4.** Unwrapped phase, corresponding to deformation 5µs after the impact

Both holograms are reconstructed separately as described in chapter 3. The interference phase map is then calculated by subtracting the reconstructed phase distributions according to Eq. (4.3).

As a typical result, the interference phase modulo $2\pi$ and the unwrapped phase are shown in figures 4.3 and 4.4. Since the sensitivity vector is nearly perpendicular to the surface, the unwrapped phase corresponds to the deformation field in z-direction 5µs after the impact.

### 4.2.2 Mechanical Materials Properties

Digital holographic displacement measurement can used to determine mechanical and thermal material properties such as the Young's modulus, the Poisson ratio and the thermal expansion coefficient. For the derivation of these quantities a procedure for the evaluation of the three-dimensional shape and the three-dimensional displacement of the object under test, a physical model of the behavior of the loaded object and the knowledge about the applied load are necessary. Such a measurement procedure based on DHI has been realized [144]. The physical model must contain one or more of the material constants as parameters. A numerical fit into the measured data according to the physical model delivers the wanted parameters within an accuracy that is determined by the numerical reliability of the model. An outline of the complete evaluation process is shown in figure 4.5 (DHI shape measurement is explained in chapter 4.3).

The calculation of the above mentioned material quantities requires to measure the whole three-dimensional displacement vector field. As stated in chapter 2.7 at least 3 interferograms of the same surface with linear independent sensitivity vectors are necessary. The interferometer consists of an optimized arrangement with 4 illumination directions and 1 observation direction to measure the 3D-displacements and coordinates precisely, figure 4.6. It includes a CCD-camera, a laser, a beam splitter cube to guide the reference beam to the CCD target and some optics for beam shaping. Optionally, a fibre coupler can be included to switch several illumination directions for varying the sensitivity vector. Such an interferometer can be designed in a very compact way, figure 4.6. Micro silicon beams are used as test samples, figure 4.7.

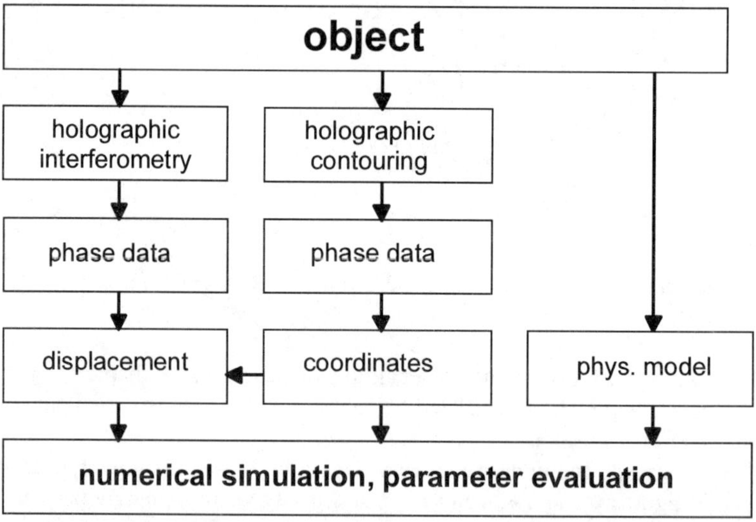


**Fig. 4.5.** Evaluation scheme for material properties

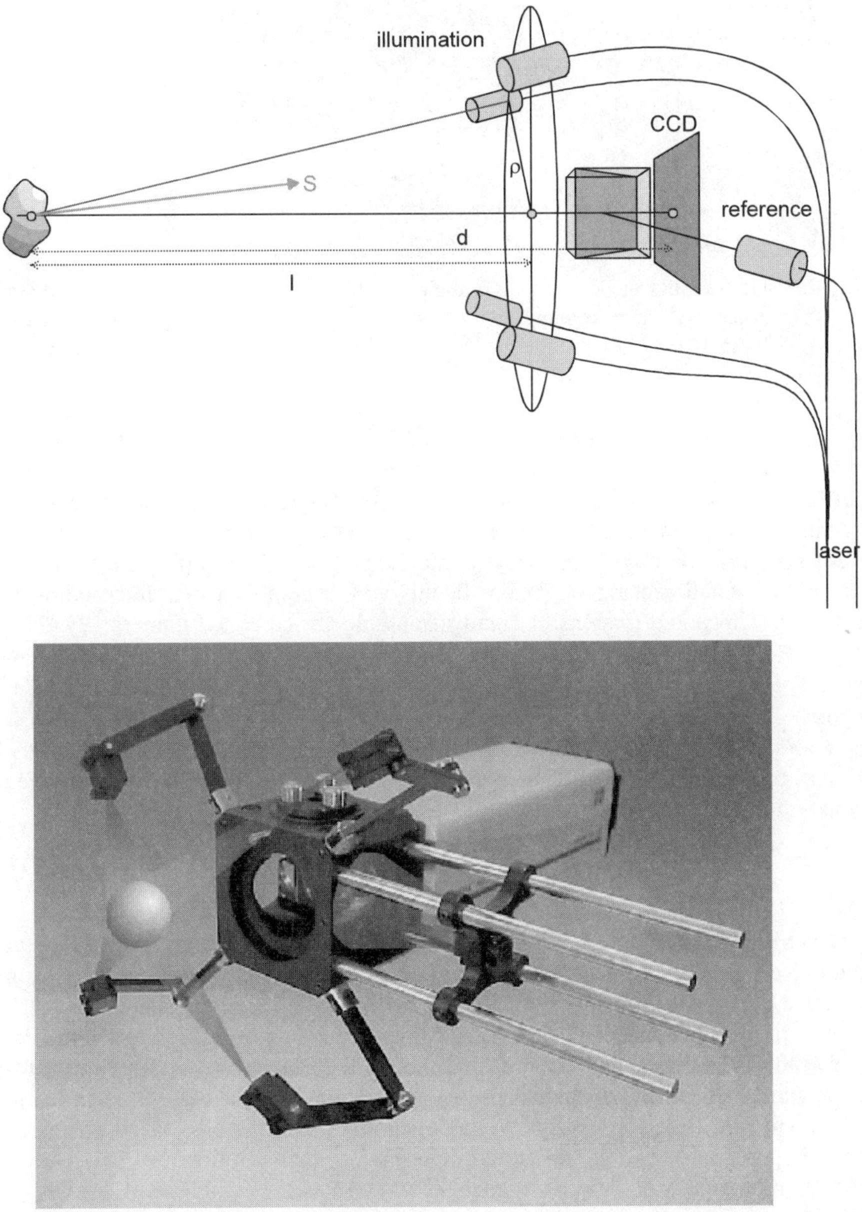


**Fig. 4.6.** DHI set-up with four illumination directions (top) and its practical implementation (bottom, photo: BIAS)

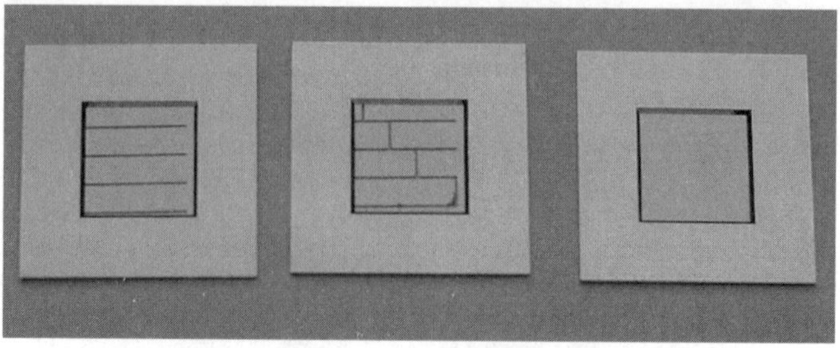

**Fig. 4.7.** Test samples made on 100mm diameter silicon wafer. The size of the quadratic structure is 9mm × 9mm, the thickness of the components is between 10μm and 40μm.

The first example is the determination of the Poisson ratio by DHI. Moments of force are applied to a rectangular sample at opposite sides. Fig. 4.8a shows a typical loading machine designed especially for small objects. The small dimensions of the samples demand a precise adjustment of all components: the "chop" which pushes against the object from above, the support and the sample which has the shape of a small rectangular beam. In this way a homogeneous deformation is achieved. Unwanted torsions of small magnitude are corrected numerically. This can be done easily with the use of the mod 2π-phase maps from Digital Holography. The resulting deformation is recorded and evaluated. The deflection causes hyperbolic structures in the 2π-phase map, figure 4.8b. Conventionally, the Poisson ratio is derived numerically from the angle between the asymptotic lines of the fringes of equal phase [170]. The deformation can be formulated by the following equation in first order approximation:

$$u(y,z) = -\frac{1}{2R}\left[z^2 + \nu\left(a^2 - y^2\right)\right] \tag{4.5}$$

Here $u$ means the deformation in x-direction at position $(y,z)$ on the surface of the object. $\nu$ stands for the Poisson ratio. $R$ is the radius of curvature and $a$ is a constant parameter. $R$ results from a slight deflection in the initial state. This helps to ensure a proper mechanical contact between the support and the sample. Eq. (4.5) shows that the upper and lower surface of the sample are deformed to parabolic curves where the inside is bent in a convex way and the outside is curved in a concave way. Since this analytical model contains the Poisson ratio as a parameter it is possible to use the measured deformation for its evaluation. This is performed numerically by approximating the model to the data (figure 4.8c) with a least-square-fit, figure 4.8d.

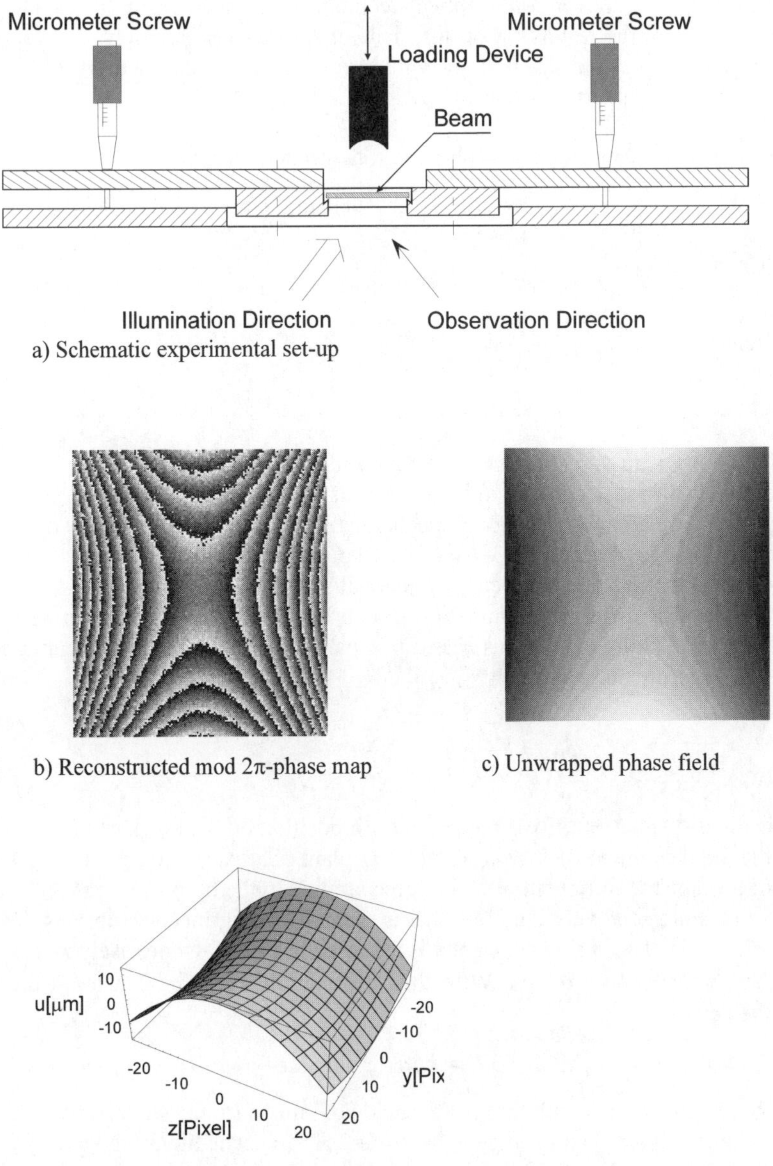


a) Schematic experimental set-up

b) Reconstructed mod $2\pi$-phase map

c) Unwrapped phase field

d) Approximated function

**Fig. 4.8.** Measurement of the Poisson ratio by DHI (from [144])

The reproducibility and accuracy of the values obtained by this method is good in comparison to conventional optical techniques for small samples. Table 4.1 contains some of the results for beams made of spring steel, structural steel and titanium. The values correlate with the values given by the manufacturers within the tolerances of the material batches.

**Table 4.1.** Measured Poisson ratios compared with literature values

| Material | Width [mm] | Thickness [mm] | Length [mm] | Poisson Ratio measured | Poisson Ratio literature |
|---|---|---|---|---|---|
| Spring steel | 1.20 | 0.20 | 12.0 | 0.288 | 0.29-0.31 |
| Spring steel | 2.00 | 0.10 | 12.0 | 0.301 | 0.29-0.31 |
| Structural steel | 1.00 | 0.50 | 10.0 | 0.338 | 0.29-0.31 |
| Structural steel | 1.50 | 0.50 | 10.0 | 0.345 | 0.29-0.31 |
| Titanium | 2.00 | 0.80 | 10.0 | 0.359 | 0.361 |
| Titanium | 1.00 | 0.80 | 10.0 | 0.381 | 0.361 |

The Young's modulus can be determined in a similar way as the Poisson ratio if the physical model contains this quantity as a parameter. Small silicon beams which are clamped at one edge and mechanically loaded at the opposite edge with a defined force are used. The 3D-surface displacement (u,v,w) (Fig. 4.9c) can be measured with the interferometer by evaluating at least 3 interferograms (Fig. 4.9b) made with different illumination directions. A model of the beam bending containing the Young's modulus $E$ as a free parameter is the basis for a numerical fit of the experimental values:

$$u(y) = \frac{Fl^3}{6EI_y}\left(2 - 3\frac{y}{l} + \frac{y^3}{l^3}\right) \tag{4.6}$$

$u$ is the displacement in x-direction and $y$ a position on the beam of the length $l$. $I_y$ is the axial moment of inertia in the $(x,z)$-plane that can be estimated with the help of a shape measurement. $F$ is the force applied to the tip of the beam. The applied forces are relatively small so that a special loading mechanism was developed, Fig. 4.9a. The spring constant $k$ is assumed to be known precisely as well as the displacement $\Delta a = a - a'$. With this information the force can be evaluated with the equation

$$F = k\Delta a \tag{4.7}$$

Several experiments with thin beams made of silicon (dimensions: length 3mm, width 1mm) delivered an average value of E=162MPa. The literature value (in the considered crystal direction) is about 166MPa. These values can vary in a big range according to the material's history, treatment and degree of impurity.

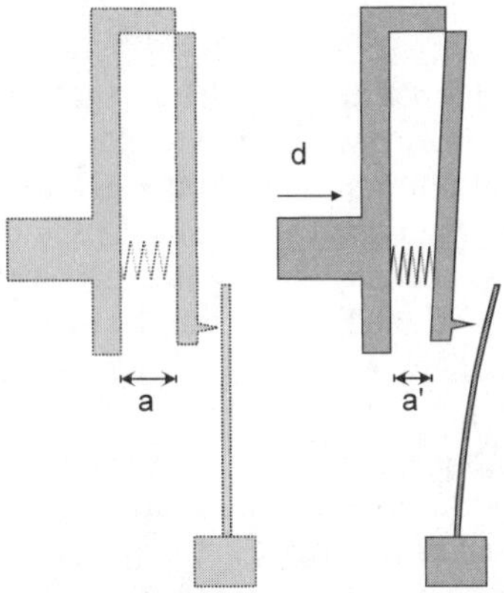


a) working principle of the loading mechanism for the small samples

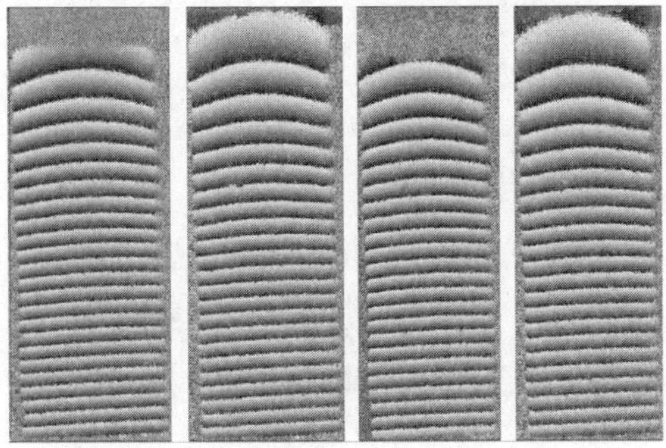

b) four $2\pi$-phase maps recorded from four different illumination directions

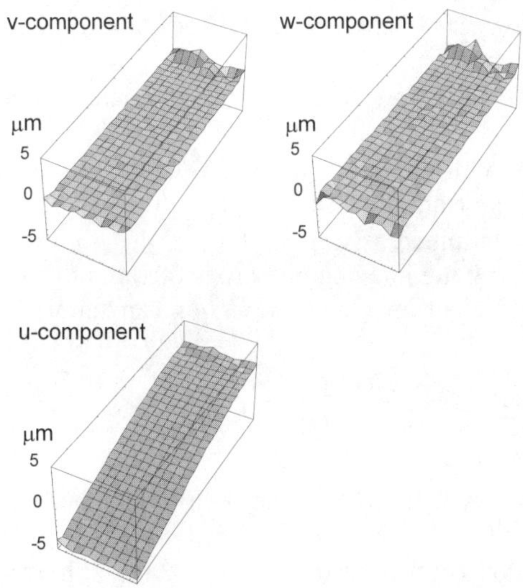


c) deformation calculated in cartesian coordinates (scale of the plots in μm)

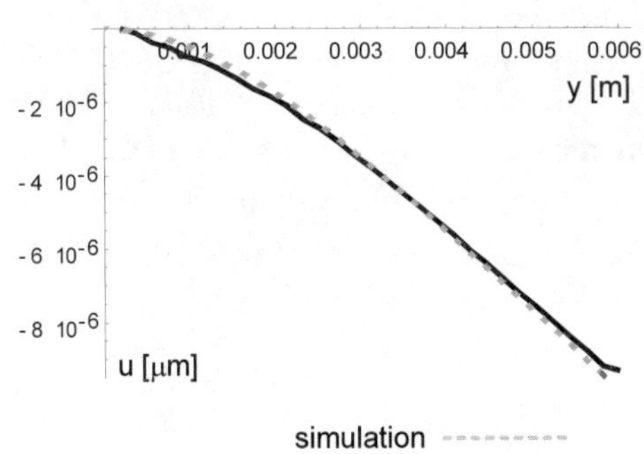


d) profile of the deformation in x-direction

**Fig. 4.9** Determination of Young's modulus by DHI

### 4.2.3 Thermal materials properties

DHI is applied also to measure thermal materials properties, e. g. the thermal expansion coefficient. For interferometric investigations of the thermal behavior it must be ensured that thermal turbulences and non-uniform temperature distributions are avoided. Therefore a vacuum chamber is used that can be supplied with adapted loading devices, figure 10a. The thermal loading device is capable of keeping a constant temperature within an accuracy of 0.02°C in a range of about 20°C up to 180°C, figure 4.10b. The digital holographic interferometer is mounted outside at the observation window of the chamber, figure 4.10c.

A mono-crystal silicon beam (figure 4.10d) with phosphor coating is used as a test object. The interferograms are recorded at various temperature differences. The complete evaluation process can be summarized as follows:

- the geometry of the setup is measured to get the geometry matrix for the evaluation of the 3 displacement components
- 4 holograms are recorded in the initial state of the object
- the object is loaded thermally and recorded holographically from four different illumination directions
- the displacement vector components $(u,v,w)$ are calculated based on the evaluation of the 4 $2\pi$-phase maps
- rigid body motions were separated from internal deformations of the object itself by subtracting the mean movement from the displacement values.
- the absolute length change $\Delta L$ is determined as well as the total length of the beam which can be performed by using the imaging properties of Digital Holography. The thermal expansion coefficient in y- and z-direction can simply be calculated by using the equation

$$\alpha = \frac{\Delta L}{L_0 \Delta T} \tag{4.8}$$

The extension in x-direction is too small to be detected with the method applied.

As an example the thermal expansion coefficient $\alpha$ of a small 2mm x 9mm x 100μm mono-crystal silicon beam has been measured. Fig. 4.11 shows the four resulting $2\pi$-phase maps. The applied temperature difference $\Delta T$ is 30°C. After elimination of the rigid body motion the three deformation components are evaluated as shown in Fig. 4.12. Considering the dimensions of the beam a value for $\alpha$ of about $\alpha = 2.92 \times 10^{-6}\,1/K$ results. The literature values vary in a big range due to different measurement methods, conditions and material batches: $\alpha = 2.4...6.7 \times 10^{-6}\,1/K$ .

a) Vacuum chamber with the supply channel

b) Equipment for thermal loading

c) Interferometer mounted on the inspection window

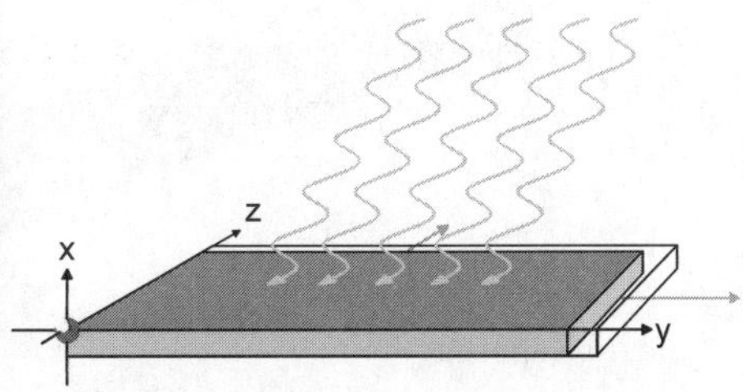


d) coordinate system used for the calculation of the thermal expansion coefficient

**Fig. 4.10.** Determination of the thermal expansion coefficient by DHI (photos: BIAS)

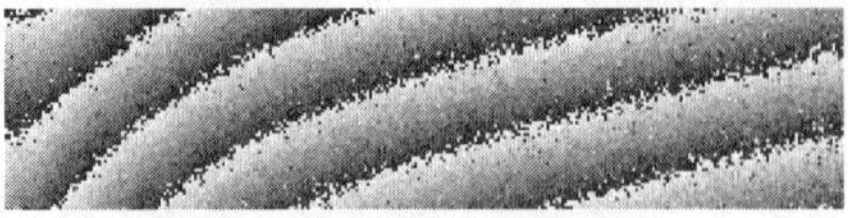

a) illumination direction 1

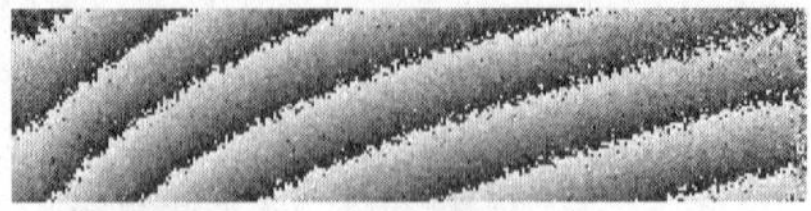

b) illumination direction 2

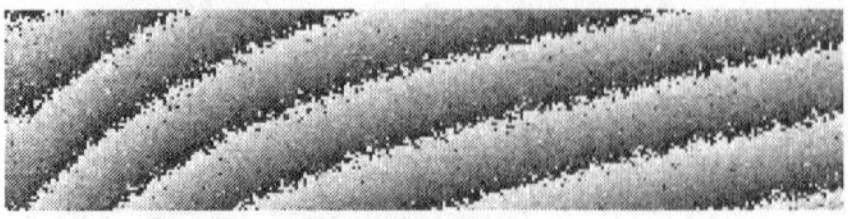

c) illumination direction 3

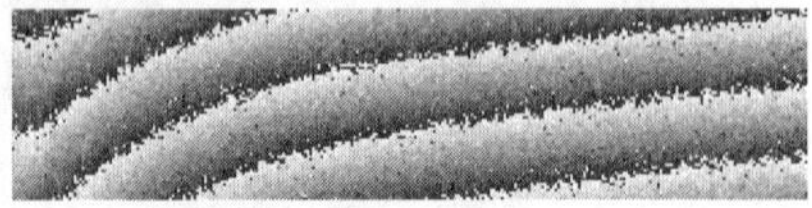

d) illumination direction 4

**Fig. 4.11.** 2π-phase maps due to deformation by thermal loading, four different illumination directions

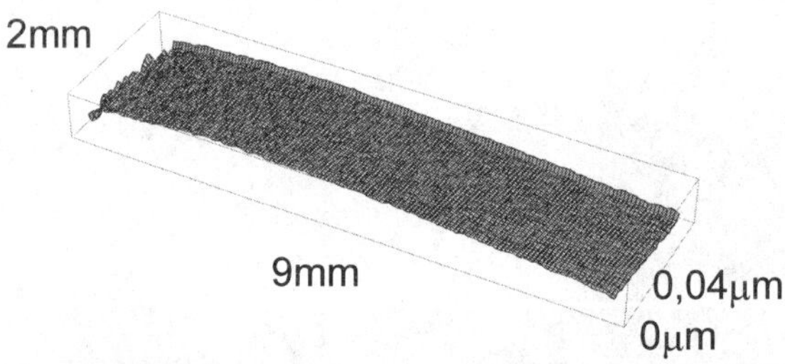


a) u-component

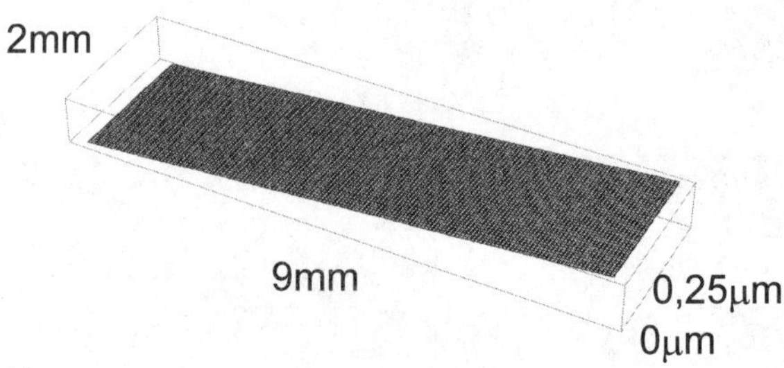


b) v-component

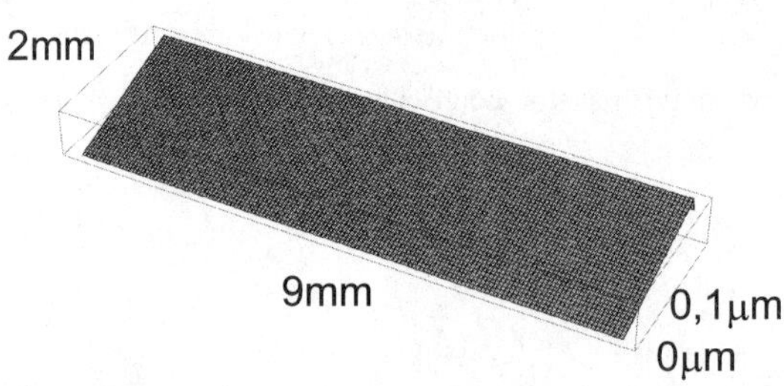


c) w-component

**Fig. 4.12.** 3D-displacement vector components (u,v,w) of thermally loaded object

### 4.2.4 Non-Destructive Testing

Non-Destructive Testing (NDT) is used to test materials and components without destruction. NDT methods are applied e. g. in aircraft industry, in power plants and in automotive production. Several physical techniques exist for NDT: Ultrasonic testing, eddy current measurements, penetrant testing, x-ray and also optical methods like HI, ESPI and shearography.

Holographic Non-Destructive Testing (HNDT) measures the deformation due to mechanical or thermal loading of a specimen. Flaws inside the material are detected as an inhomogeneity in the fringe pattern corresponding to the surface deformation.

HNDT can be used wherever the presence of a structural weakness results in a surface deformation of the stressed component. The load can be realized by the application of a mechanical force or by a change in pressure or temperature. Holographic NDT indicates deformations down to the submicrometer range, so loading amplitudes far below any damage threshold are sufficient to produce detectable fringe patterns.

In HNDT it is sufficient to have one fringe pattern of the surface under investigation. Quantitative evaluation of the displacement vector field is usually not required. The fringe pattern is evaluated qualitatively by the human observer or, more and more, by fringe analysis computer codes. Irregularities in the interference pattern are indicators of flaws within the component under investigation.

As discussed in the preceding chapters, DHI generates not a fringe pattern, but directly the interference phase map. This phase map is therefore used for flaw detection in Digital Holographic Non-Destructive Testing (DHI NDT).

As an application for DHI NDT testing of a pressure vessel [138] is described. Such vessels are used as gas tanks in satellites, see figure 4.13. The diameter is in the order of 1 or 2 meter and the thickness of the wall in only about 1 millimetre. Typical flaws to be detected are cracks or reduced thickness of the wall.

The surface of the tank is divided into segments of about $5cm \times 5cm$. For each segment, a series of digital holograms is recorded. Between the exposures, the pressure inside the tank is varied by a few hundred hPa. As a typical result the interference phase between one pair of holograms is shown in figure 4.14. The disturbance in the middle is an indication for a flaw. The interference phase can also be converted into a continuous phase by unwrapping the 2π-jumps. However, for flaw detection the unwrapped phase map is often more suitable.

In conventional holographic NDT the load applied to the object under investigation is restricted to certain limit. Loads above this limit lead to irresolvable fringe densities. In the numerical reconstruction process of DHI, the phase difference between any pair of exposures can be calculated. Even if the total deformation between the first and the last hologram is too high for direct evaluation, the total phase difference can be calculated step by step as the sum of the individual phase changes:

$$\Delta\varphi_{total} = \Delta\varphi_{1\to 2} + \Delta\varphi_{2\to 3} + \Delta\varphi_{3\to 4} + \ldots \tag{4.9}$$

However, a drawback of DHI compared to conventional HI shall be emphasized: For visual flaw detection it is sometimes of advantage to vary the observation direction continuously (dynamic evaluation). This is possible for holograms recorded on a photographic plate with dimensions of about $10cm \times 10cm$ or more, but until now not for digital holograms recorded by a CCD with only about $1cm \times 1cm$ area. However, future progress in camera technology and computer hardware may solve this problem.

**Fig. 4.13.** Satellite tank

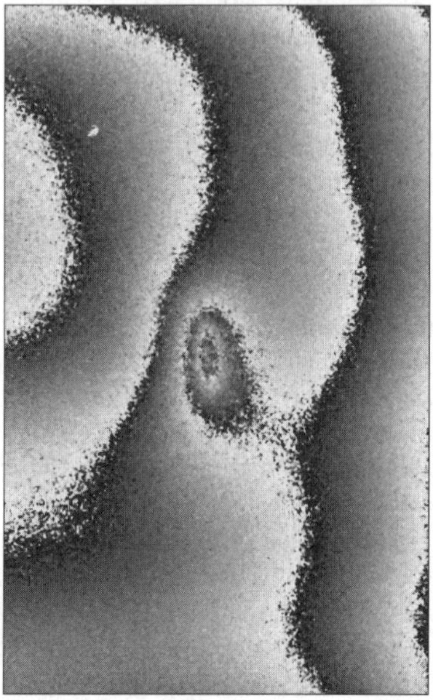

**Fig. 4.14.** Non-destructive testing by DHI

## 4.3 Shape Measurement

### 4.3.1 Two-Illumination-Point Method

The two contouring techniques discussed in chapter 2.7.3 for conventional HI are also applied in DHI.

For the Two-Illumination-Point method it is necessary to record two digital holograms of the same surface, but with different object illumination points. Both holograms are reconstructed separately. The interference phase map, which represents the object shape, is then calculated by subtracting the individual phase distributions according to Eq. (4.3). The result is a wrapped phase map, which is interpreted similar to the contour fringe pattern discussed in chapter 2.7.3. The phase change between two adjacent 2π-jumps is

$$\Delta\varphi = \frac{2\pi}{\lambda}\vec{p}\vec{s} \tag{4.10}$$

with $\vec{p}$ and $\vec{s}$ as defined previously. By analogy with Eq. (2.96) the distance between two adjacent 2π-jumps is

$$\Delta H = \frac{\lambda}{2\sin\frac{\theta}{2}} \tag{4.11}$$

where $\theta$ is the angle between the two illumination directions.

DHI two-illumination-point contouring can be carried out e. g. with the set-up depicted in figure 4.15. Optical fibres are preferably used to guide the illumination wave. The fibre output is the illumination source point. The shift is realized e. g. by a motor driven translation stage. The first digital hologram is recorded with the fibre end at position S. For the second hologram the fiber is shifted slightly to position S'.

In order to obtain a maximum sensitivity in a direction normal to the surface, the illumination should come from the side, i. e. the angle between illumination direction and observation direction is near 90°. Still, a such flat illumination directions may cause shadows due to surface variations. The optimum illumination direction is therefore always a compromise between maximum sensitivity and minimum shadows in the reconstructed images.

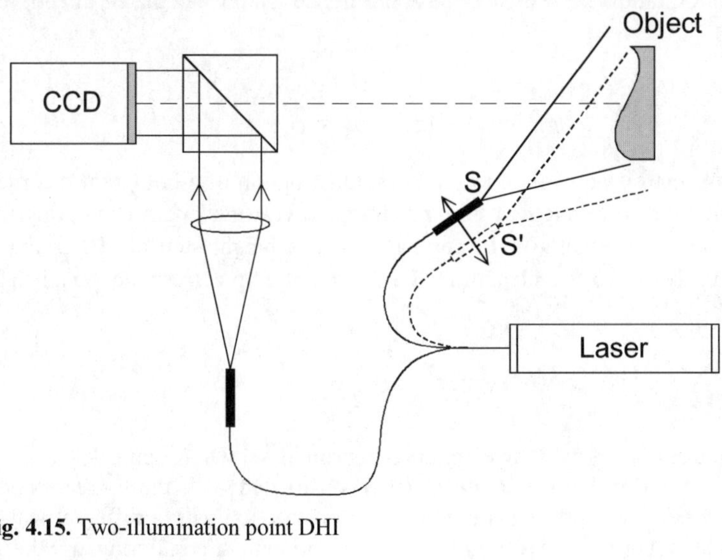


**Fig. 4.15.** Two-illumination point DHI

### 4.3.2 Two- and Multi-Wavelength Method

For shape measurement by the two-wavelength method two holograms are recorded with different wavelengths $\lambda_1$ and $\lambda_2$. In conventional HI both holograms are recorded on a single photographic plate. Both holograms are reconstructed by the same wavelength, e. g. $\lambda_1$. That is why two images of the object are generated. The image recorded and reconstructed by $\lambda_1$ is an exact duplicate of the original object surface. The image which has been recorded with $\lambda_1$, but is reconstructed with $\lambda_2$ is slightly shifted in observation direction (towards the observer, see imaging equations in chapter 2.6.2) with respect to the original surface. These two reconstructed images interfere.

The concept of two-wavelength contouring has been introduced into Digital Holography, too [62, 129, 143]. Two holograms are recorded with $\lambda_1$ and $\lambda_2$ and stored electronically, e. g. with the set-up depicted in figure 4.16. In contrast to conventional HI using photographic plates, both holograms can be reconstructed separately by the correct wavelengths according to the theory of chapter 3. From the resulting complex amplitudes $\Gamma_{\lambda 1}(\xi,\eta)$ and $\Gamma_{\lambda 2}(\xi,\eta)$ the phases are calculated:

$$\varphi_{\lambda 1}(\xi,\eta) = \arctan\frac{\operatorname{Im}\Gamma_{\lambda 1}(\xi,\eta)}{\operatorname{Re}\Gamma_{\lambda 1}(\xi,\eta)} \tag{4.12}$$

$$\varphi_{\lambda 2}(\xi,\eta) = \arctan\frac{\operatorname{Im}\Gamma_{\lambda 2}(\xi,\eta)}{\operatorname{Re}\Gamma_{\lambda 2}(\xi,\eta)} \tag{4.13}$$

As for deformation analysis the phase difference is now calculated directly by subtraction:

$$\Delta\varphi = \begin{cases} \varphi_{\lambda 1} - \varphi_{\lambda 2} & if \varphi_{\lambda 1} \geq \varphi_{\lambda 2} \\ \varphi_{\lambda 1} - \varphi_{\lambda 2} + 2\pi & if \varphi_{\lambda 1} < \varphi_{\lambda 2} \end{cases} \tag{4.14}$$

This phase map is equivalent to the phase distribution of a hologram recorded with the synthetic wavelength $\Lambda$. In conventional two-wavelength contouring the distance between adjacent fringes corresponds to a height step of $\Lambda/2$, see Eq. (2.90). Similarly in two-wavelength DHI a $2\pi$ phase jump corresponds to a height step of $\Lambda/2$:

$$\Delta H = \frac{\lambda_1 \lambda_2}{2\left|\lambda_1 - \lambda_2\right|} = \frac{\Lambda}{2} \tag{4.15}$$

A typical example of two-wavelength contouring is shown in figure 4.17.

In DHI contouring both holograms are reconstructed with the correct wavelength. Distortions resulting from hologram reconstruction with a different wavelength than the recording wavelength, as in conventional HI contouring, are therefore avoided.

A modified contouring approach, which is referred to as *multiwavelength contouring*, has to use more than two illumination wavelengths to eliminate ambiguities inherent to modulo 2π phase distributions [66, 67, 110, 163]. The advantage of this technique is that it can also be used with objects that have phase steps or isolated object areas.

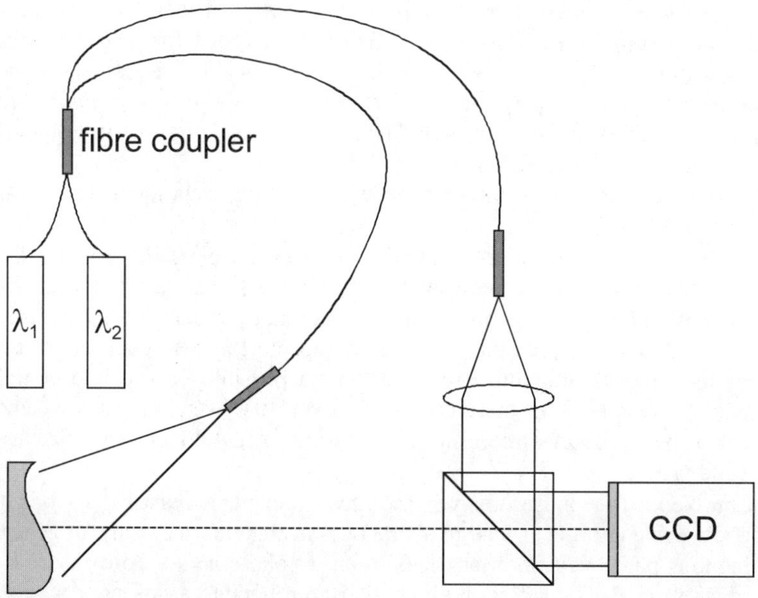


**Fig. 4.16.** Two-wavelength DHI

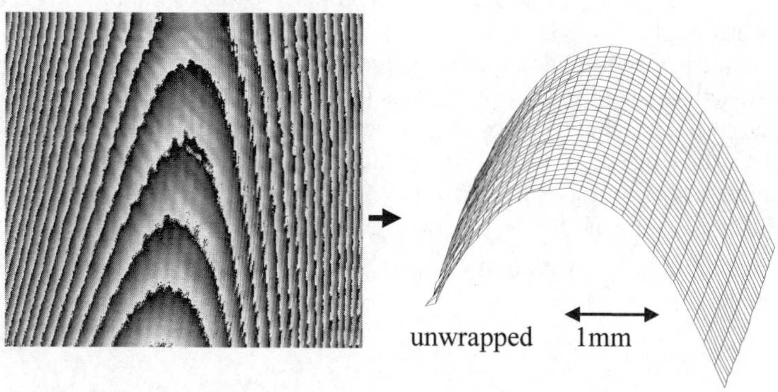


**Fig. 4.17.** Shape registration of a cylindrical lens by two-wavelength contouring. Visible part: 3mm × 2mm

### 4.3.3 Hierarchical Phase Unwrapping

As stated earlier the process of phase unwrapping is always the same for conventional HI as well as for DHI and in general also for all methods which generate modulo 2π images. In chapter 2.7.6 a simple phase unwrapping method is described. However, for DHI multiwavelength contouring a special unwrapping procedure named *hierarchical phase unwrapping* is useful. The basic idea of this technique was originally developed by Nadeborn and Osten for projected fringe techniques with incoherent light [99, 107], but it is applicable for all measurement techniques which generate periodic data. *Hierarchical phase unwrapping* is particularly well suited in conjunction with DHI multi-wavelength contouring. The technique is discussed in this context.

The practical application of interferometric contouring techniques leads to following problems [142]:

- Fringe counting problem: The interference phases are periodic. For the Two-Wavelength method the periodic length is half the synthetic wavelength, see Eq. (4.15). If edges, holes, jumps or other discontinuations are on the measured surface, it is often not possible to count the absolute interference order, resp. phase value. An unambiguous evaluation is not possible. In order to generate unambiguous results, it is therefore necessary to use a synthetic wavelength bigger than twice the maximum height variation of the object. But this causes the second problem:
- Noise problem: The phase noise of the measurement increases the larger the synthetic wavelength gets. A large synthetic wavelength, necessary to generate unambiguous phase values, often leads to larger phase noise. To measure e. g. the surface profile of an object with about 10mm height variations, a synthetic wavelength of at least $\Lambda = 20mm$ is necessary. In practice the phase noise limits the achievable measurement resolution to about 1/10 of the synthetic wavelength; i. e. the accuracy of the measured height data is only 0.2mm. This is too low for many applications.

The basic idea of hierarchical phase unwrapping is to start with a large synthetic wavelength to avoid phase ambiguities [163]. This measurement is not very accurate due to noise. The accuracy is now improved by systematic reduction of the synthetic wavelengths, while the information of the preceding measurements is used to eliminate ambiguities.

The procedure starts with a synthetic wavelength $\Lambda_1$, which is larger than twice the maximum height variation of the object. The height at a certain position is then given by

$$z_1 = \frac{\Lambda_1}{2} \frac{\Delta\varphi_1}{2\pi} \tag{4.16}$$

where $\Delta\varphi_1$ is the measured interference phase at this position. This result is unambiguous, but strongly disturbed by noise. Now the synthetic wavelength is reduced to $\Lambda_2$ and a new phase measurement is made. The resulting height coordinate

$$\hat{z}_2 = \frac{\Lambda_2}{2}\frac{\Delta\varphi_2}{2\pi} \tag{4.17}$$

is not unambiguous, this is indicated by the "^". In the next step the difference between the result of the first measurement $z_1$ and the second measurement $\hat{z}_2$ is calculated:

$$\Delta z = z_1 - \hat{z}_2 \tag{4.18}$$

Furthermore it is calculated which multiple of the periodic length $\Lambda_2/2$ is contained in the difference $\Delta z$ (rounded number):

$$N = floor\left(2\frac{\Delta z}{\Lambda_2} + 0.5\right) \tag{4.19}$$

The function $f(x) = floor(x)$ delivers the maximum integer value, which is smaller than *x*. The correct result of the second measurement is now:

$$z_2 = \hat{z}_2 + \frac{\Lambda_2}{2}N \tag{4.20}$$

This result is unambiguous as well as $z_1$, but it has a better accuracy compared to $z_1$ due to the smaller synthetic wavelength. The procedure is continued with smaller wavelengths as long as the resolution is below a desired value. After *n* iterations one gets the formula:

$$z_n = \hat{z}_n + \frac{\Lambda_n}{2} floor\left(2\frac{z_{n-1} - \hat{z}_n}{\Lambda_n} + 0.5\right) \tag{4.21}$$

In practice the number of measurements to reach the desired resolution should be as small as possible. This minimum or optimum number depends on the noise. Let $\varepsilon_n$ be the inaccuracy in a measurement using the synthetic wavelength $\Lambda_n$. The true height coordinate $z_{true}$ lies within an interval limited by

$$z_{\max} = z_{meas} + \frac{\varepsilon_n}{2}\frac{\Lambda_n}{2} \tag{4.22}$$

$$z_{\min} = z_{meas} - \frac{\varepsilon_n}{2}\frac{\Lambda_n}{2}$$

where $z_{meas}$ is the value determined by the measurement. Then the next measurement with $\Lambda_{n+1}$ takes place. The inaccuracy of this measurement is given by

$\varepsilon_{n+1}$. Now the noise of the n$^{th}$ and the (n+1)$^{th}$ measurement is considered for estimating the interval limits of the true height coordinate:

$$z_{max} = z_{meas} + \frac{\varepsilon_n}{2}\frac{\Lambda_n}{2} + \frac{\varepsilon_{n+1}}{2}\frac{\Lambda_{n+1}}{2}$$
$$z_{min} = z_{meas} - \frac{\varepsilon_n}{2}\frac{\Lambda_n}{2} - \frac{\varepsilon_{n+1}}{2}\frac{\Lambda_{n+1}}{2} \tag{4.23}$$

The difference between the maximum possible and the minimum possible height coordinate is:

$$\left| z_{max} - z_{min} \right| = \varepsilon_n \frac{\Lambda_n}{2} + \varepsilon_{n+1}\frac{\Lambda_{n+1}}{2} \tag{4.24}$$

A correct recovery of the absolute height coordinate is only possible, if following condition is satisfied for the (n+1)$^{th}$ measurement:

$$\frac{\Lambda_{n+1}}{2} \geq \left| z_{max} - z_{min} \right| \tag{4.25}$$

A smaller value for $\Lambda_{n+1}/2$ than $\left| z_{max} - z_{min} \right|$ would lead to ambiguities, because $\hat{z}_{n+1} + N\Lambda_{n+1}/2$ as well as $\hat{z}_{n+1} + (N+1)\Lambda_{n+1}/2$ are possible height values within the interval limits. The optimum period (half synthetic wavelength) is achieved for the equals sign.

The combination of Eq. (4.24) and (4.25) with the equals sign results to:

$$\Lambda_{n+1} = \Lambda_n \frac{\varepsilon_n}{1-\varepsilon_{n+1}} \tag{4.26}$$

This condition determines the optimum choice of the sequence of synthetic wavelengths depending on each measurement's accuracy.

## 4.4 Measurement of Refractive Index Variations

Digital HI is also used to measure refractive index variations within transparent media, e. g. with the set-up of figure 4.18. The expanded laser beam is divided into reference and object beam. The object beam passes the transparent phase object and illuminates the CCD. The reference beam impinges directly on the CCD. Both beams interfere and the hologram is digitally recorded. The set-up is very similar to a conventional Mach-Zehnder interferometer. The difference is that the interference figure here is interpreted as a hologram, which can be reconstructed with the theory of chapter 3. Therefore all features of Digital Holography like direct access to the phase or numerical re-focussing are available.

Just like for deformation analysis two digital holograms are recorded: The first exposure takes place before, and the second after the refractive index change.

These digital holograms are reconstructed numerically. From the resulting complex amplitudes $\Gamma_1(\xi,\eta)$ and $\Gamma_2(\xi,\eta)$ the phases are calculated by Eq. (4.1) and (4.2). Finally the interference phase is calculated by subtraction according to Eq. (4.3).

In the reconstruction of holograms recorded by the set-up of figure 4.18 the undiffracted reference wave, the real image and the virtual image are lying in one line. The images overlap, which causes distortions. The undiffracted reference wave can be suppressed by filtering with the methods discussed in chapter 3.3.1. The overlapping of the unwanted twin image (either the virtual image if one focuses on the real image or vice versa) can be avoided by slightly tilting the reference wave, as discussed in chapter 3.3.2. In this case the images are spatially separated.

The interferometer of figure 4.18 is sensitive to local disturbances due to imperfections in optical components or dust particles. The influence of these disturbances can be minimized if a diffusing screen is placed in front of or behind the phase object. In this case the unfocused twin image appears only as a diffuse background in the images, which does not disturb the evaluation. If a diffuser is introduced no additional tilting of the reference wave is necessary for image separation. A disadvantage of using a diffuser is the generation of speckles due to the rough surface.

In figure 4.19 a typical interference modulo 2 π image of a transparent phase object is shown. The holograms are recorded with the set-up of figure 4.18 (without diffuser). The object volume consists of a droplet of toluene, which is introduced into the liquid phase water/ acetone. The refractive index changes are caused by a concentration gradient, which is induced by the mass transfer of acetone into the droplet.

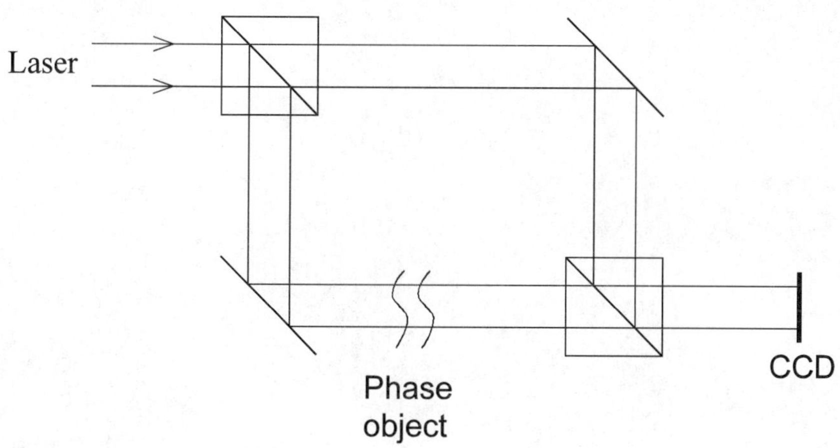


**Fig. 4.18.** DHI set-up for transparent phase objects

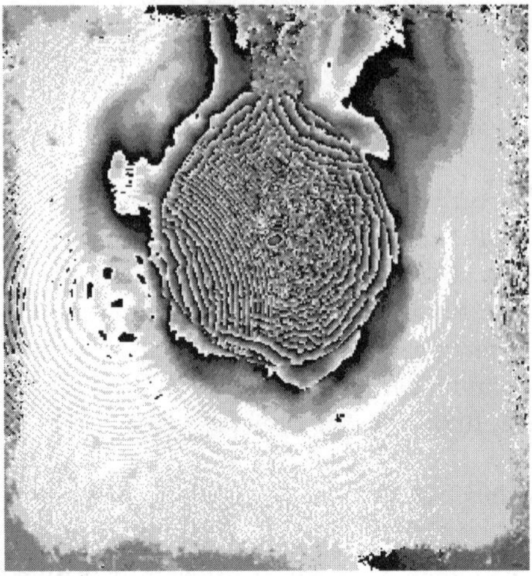

**Fig. 4.19.** Wrapped phase of a liquid system

# 5 Digital Holographic Microscopy

## 5.1 Direct Method

The depth of field of imaging systems decreases with increasing magnification. In microscopy the depth of field is therefore very limited due to the high magnification. The investigation of a three dimensional object with microscopic resolution requires therefore certain refocusing steps. Digital Holography offers the possibility to focus on different object layers by numerical methods. In addition, the images are free of aberrations due to imperfections of optical lenses. Fundamental work in the field of Digital Holographic Microscopy (DHM) has been done by Haddad et al. [46], Kreuzer and Pawlitzek [80], Kebbel, Hartmann and Jüptner [65] and by Coppola et. al. [16, 17].

In order to obtain a high lateral resolution in the reconstructed image the object has to be placed near to the CCD. The necessary distance to obtain a resolution $\Delta\xi'$ with the Fresnel approximation can be estimated with Eq. (3.23):

$$\Delta\xi' = \frac{\lambda d'}{N\Delta x} \quad ; \quad \Delta\eta' = \frac{\lambda d'}{N\Delta y} \tag{5.1}$$

The apostrophe is introduced in order to decide between object distance $d$ in the recording process and reconstruction distance $d'$. We will see that these distances are different for holographic microscopy. With a pixel size of $\Delta x = 10\mu m$, a wavelength of $\lambda = 500nm$, $1000\times1000$ pixels and a required resolution of $\Delta\xi' = 1\mu m$ a reconstruction distance of $d' = 2cm$ results. At such short distances the Fresnel approximation is no longer valid. The convolution approach has to be applied. On the other hand the resolution of an image calculated by this approach is determined by the pixel size of the CCD, see Eq. (3.33). Typical pixel sizes for high resolution cameras are in the range of $10\mu m \times 10\mu m$, too low for microscopy. Therefore the reconstruction procedure has to be modified.

The lateral magnification of the holographic reconstruction can be derived from the holographic imaging equations, see chapter 2.6.2. According to Eq. (2.70) the lateral magnification of the reconstructed virtual image is:

$$M = \left[1 + \frac{d}{d_r'}\frac{\lambda_1}{\lambda_2} - \frac{d}{d_r}\right]^{-1} \tag{5.2}$$

where $d_r$ and $d_r'$ describe the distances between the source point of a spherical reference wave and the hologram plane in the recording and reconstruction process, respectively. $\lambda_1$ and $\lambda_2$ are the wavelengths for recording and reconstruction. The reconstruction distance $d'$, i. e. the position of the reconstructed image, can be calculated with Eq. (2.66):

$$d' = \left[\frac{1}{d_r'} + \frac{\lambda_2}{\lambda_1}\frac{1}{d} - \frac{1}{d_r}\frac{\lambda_2}{\lambda_1}\right]^{-1} \tag{5.3}$$

If the same reference wavefront is used for recording and reconstruction it follows $d' = d$. Note that $d$, $d'$, $dr$ and $dr'$ are always counted positive in this book.

Magnification can be introduced by changing the wavelength or the position of the source point of the reference wave in the reconstruction process. In Digital Holography the magnification can be easily introduced by changing the reference wave source point. If the desired magnification factor is determined, the reconstruction distance can be calculated by combination of Eq. (5.2) and (5.3) with $\lambda_1 = \lambda_2$:

$$d' = d \cdot M \tag{5.4}$$

To enlarge the image the source point of the reference wave needs to be placed at the distance

$$d_r' = \left[\frac{1}{d'} - \frac{1}{d} + \frac{1}{d_r}\right]^{-1} \tag{5.5}$$

The reference wave is now described by

$$E_R(x,y) = \exp\left(-i\frac{2\pi}{\lambda}\sqrt{d_r'^2 + (x - x_r')^2 + (y - y_r')^2}\right) \tag{5.6}$$

where $(x_r', y_r', -d_r')$ is the position of the reference source point in the reconstruction process. The entire process is summarized as follows: After determination of the desired magnification the reconstruction distance $d'$ is calculated with Eq. (5.4). Secondly the source point position $d_r'$ of the spherical reference wave for reconstruction is calculated by Eq. (5.5).

A set-up for digital holographic microscopy is shown in figure 5.1. The object is illuminated in transmission and the spherical reference wave is coupled into the set-up via a semi-transparent mirror. Reference and object wave are guided via optical fibres. For weak scattering objects one can block the external reference wave and work with an in-line configuration.

A digital hologram of a test target is shown in figure 5.2. The corresponding intensity reconstruction is depicted in figure 5.3. The resolution is about 2.2 µm.

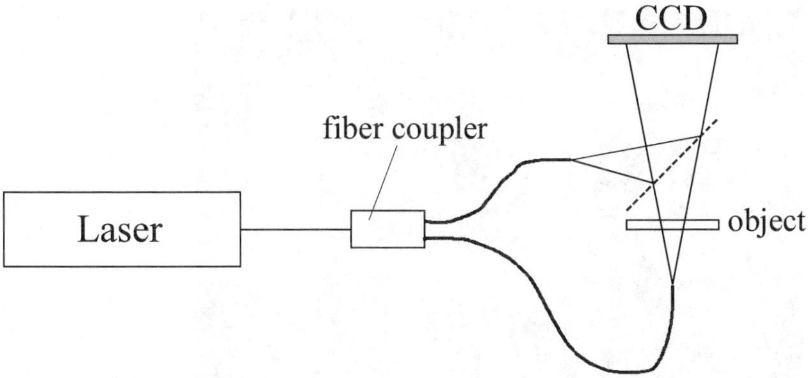


**Fig. 5.1.** Digital holographic microscope

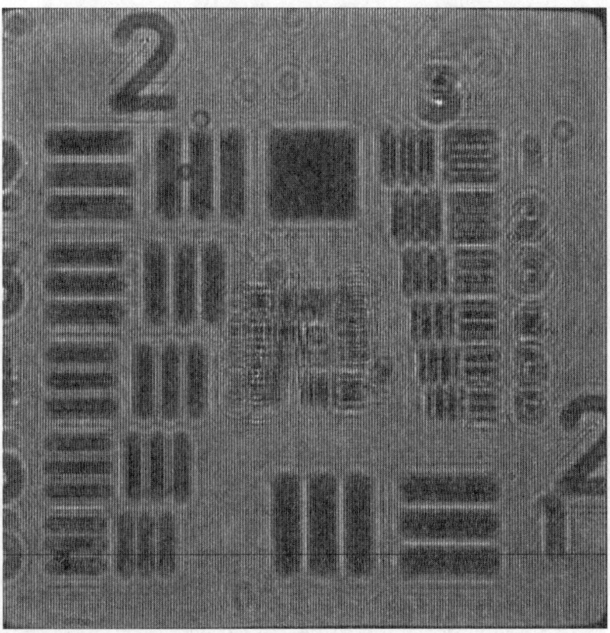

**Fig. 5.2.** Digital hologram

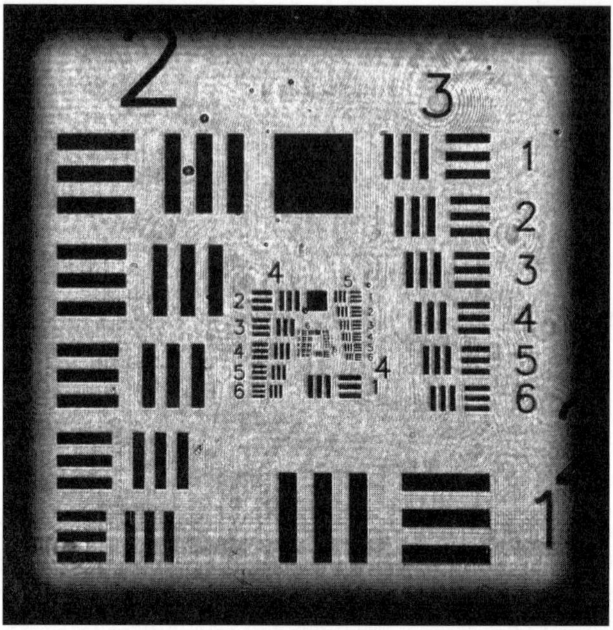

Fig. 5.3. Numerical reconstruction with microscopic resolution

## 5.2 Phase Shifting Digital Holographic Microscopy

Phase Shifting Digital Holography has been also applied to microscopy [172, 178, 179]. The principle of this method is shown in the set-up of figure 5.4. A light beam is coupled into a Mach-Zehnder interferometer. The sample to be investigated (object) is mounted in one arm of the interferometer. It is imaged onto the CCD target by a microscope objective (MO). A second objective is mounted in the reference arm in order to form a reference wavefront with the same curvature. Both partial waves interfere at the CCD target. An image of the sample superimposed by a coherent background (reference wave) is formed onto the CCD target.

A set of phase shifted images is recorded. The phase shift is realized by a piezo electric transducer (PZT) in the reference arm of the interferometer. From these phase shifted images the complex amplitude of the object wavefront in the image plane can be calculated as described in chapter 3.3.3. Numerical refocusing into any other object plane is now possible with the Fresnel-Kirchhoff integral.

The quality of images recorded with coherent light is in general worse than those recorded with incoherent light due to coherent noise. Dubois, Joannes and Legros developed therefore a phase shifting digital holographic microscope with an ordinary LED as light source [30]. The image quality improves (less speckle noise) due to the reduced spatial coherence of that light source compared to images generated by a laser.

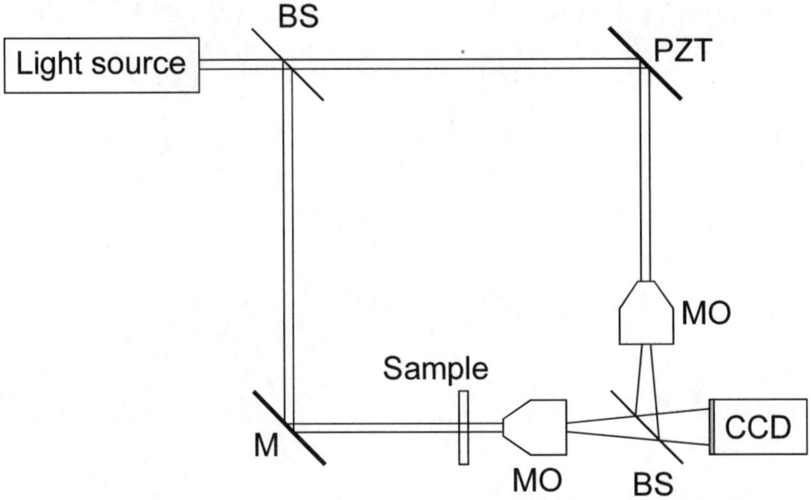


**Fig. 5.4.** Phase shifting digital holographic microscope

# 6 Special Techniques and Applications

## 6.1 Applications using Short Coherence Length Light

### 6.1.1 Light-in-Flight Measurements

Holographic recording of Light-in-Flight (LiF) was first proposed by Abramson [1-4]. He pointed out that a hologram can only image the distances in space where the optical path of the reference wave matches that of the object wave. The basic idea of LiF consists of recording a hologram of a plane object with streaking illumination by a short coherence length laser, figure 6.1. For this purpose a cw Ar-Ion laser without intracavity etalon is used. The coherence length of such laser is in the range of few millimeters or less. Alternatively also a picosecond pulsed dye laser can be used. The reference wave is guided nearly parallel to the holographic plate. In this way, only those parts of the object are recorded (and later reconstructed), for which the optical path difference (OPD) between object and reference wave is smaller than the coherence length of the light source. By changing the observation point in the developed plate, the above condition is met for different parts of the object, thus allowing observation of a wavefront as it evolves over the object.

Digital Holography has been applied to LiF recordings by Pomarico, Schnars, Hartmann and Jüptner [61, 126]. The general setup of figure 3.1 can be used for recording LiF if a short coherence length light source is used for illuminating the object. In the following investigations are described, in which an Ar-Ion laser pumped cw dye laser (Rhodamine 6G, $\lambda$ = 574 nm) is used as light source. No frequency selecting elements are installed in the laser resonator. Therefore the output spectrum consists of many oscillating modes, resulting in a coherence length, which is determined by the LiF experiments to be 2.3 mm.

The laser beam is divided into a plane reference wave illuminating the CCD array and into a diverging wave illuminating the object, figure 6.2. The path differences are provided by glass plates with different but known thicknesses. The object consists of a plane aluminum plate of 2 cm × 2 cm area, the distance between object and CCD sensor is set to 1.67 m, and the angle of illumination $\alpha$ (referred to the normal of the object) is about 80 degrees. Furthermore, a wavelength of $\lambda$ = 574nm is used and the maximum angle between object and reference wave is $q_{max}$ = 2°. In this experiment a KODAK MEGAPLUS 4.2 camera with 2048 × 2048 light sensitive elements is used as recording medium.

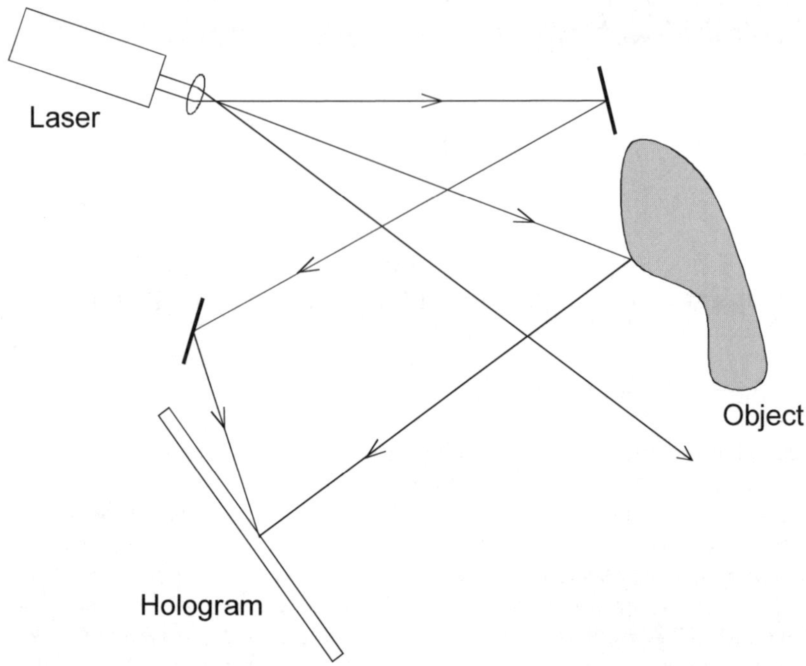


**Fig. 6.1.** Light-in-flight holography

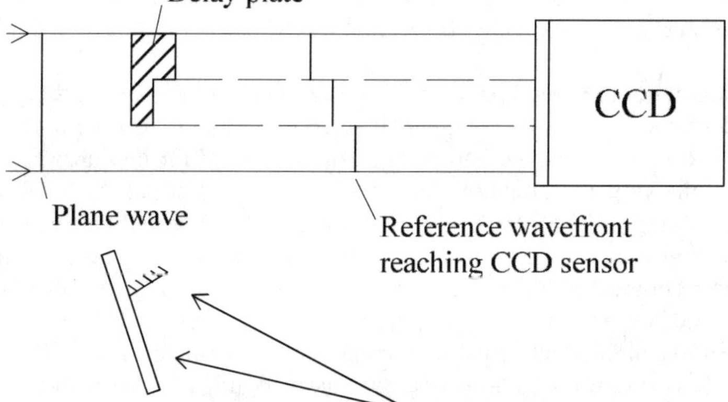


**Fig. 6.2.** Optical set-up for Digital LiF recording with delay lines by glass plates

The first experiment is performed without the glass plates in the reference arm of the interferometer. Since the object is illuminated under an angle only a part of the object could be illuminated in coherence to the reference wave. The numerical reconstruction of such a digitally recorded hologram shows, as expected, a bright stripe representing the wavefront, figure 6.3.

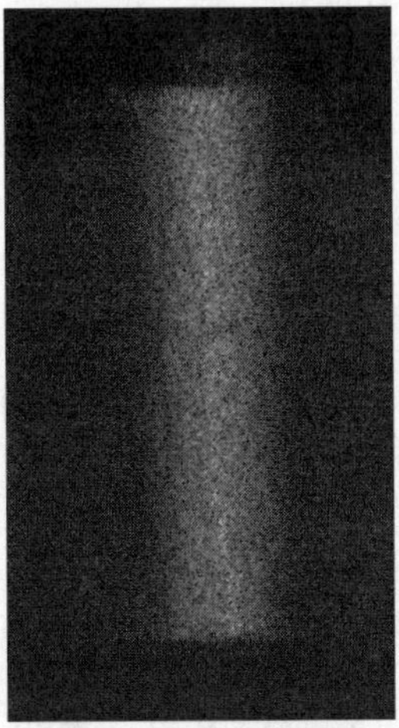

**Fig. 6.3.** Numerically reconstructed wavefront

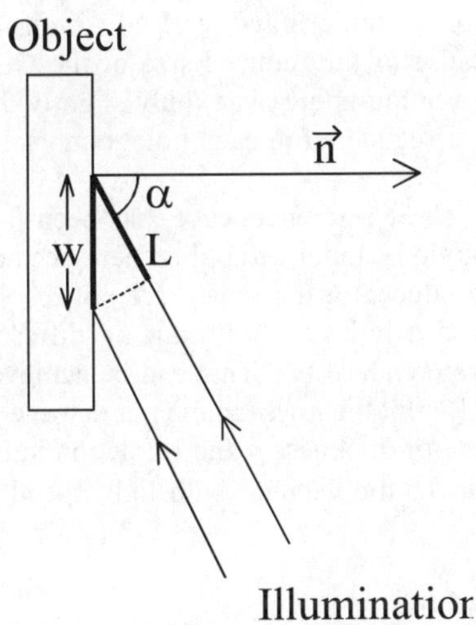


**Fig. 6.4.** Geometrical considerations for calculating the coherence length

The reconstructed image is available in a digital form and further processing is easy to be done. For example, the coherence length can be calculated from this image, figure 6.4. The width of the bright stripe (wavefront) seen at the object is determined from both, the coherence length of the light source, *L*, and the geometrical conditions of the holographic set-up. If a plane reference wave is used and the angle between the interfering waves is small ( $q_{max} = 2°$ in this example), only changes in the optical path due to the illumination beam have to be considered. In this case the bright zone at the object has a width $w$ given by

$$w = \frac{c\tau}{\sin\alpha} = \frac{L}{\sin\alpha} \tag{6.1}$$

where $\alpha$ is defined in figure 6.4 and $\tau$ is the coherence time. After measuring $w$, Eq.(6.1) can be used to calculate the coherence length $L$ of the light source using the known angle $\alpha$ of the incident wave. As the measurements of the width are disturbed by electronic and coherent noise, direct measurement of the intensity profile leads to errors. A good result can be achieved by low-pass filtering of the image and by applying the autocorrelation function to the intensity profile line. The width of the wavefront measured by this procedure values to 45 pixels. The experimental conditions are:

$\Delta x = 9\mu m$ ; $d = 1.67m$ ; $\lambda = 574nm$ ; $\alpha = 80°$

The resulting coherence length is therefore

$$L = w\sin\alpha = 45 \cdot \Delta\xi \sin\alpha = 2.3mm \tag{6.2}$$

where Eq. (3.23) is used for calculating the image resolution $\Delta\xi$.

It is also possible to apply Digital Holography to follow the evolution of a wave front in its “flight” over an object, as proposed in the original work of Abramson for conventional holography. However, because of the reduced size of the CCD target and the lower resolution compared to a holographic plate, only slightly different points of view of the wave front can be registered in each hologram. Some view-time expansion is also needed.

A possible setup for this purpose, using a skew reference wave, has been proposed by Pettersson et.al. [125]. Yet, this solution cannot be applied here because the high spatial frequencies that would be produced at the sensor are not resolvable. A solution to this problem is to record a hologram introducing different phase delays in different parts of the plane reference wave. That can be achieved e.g. by introducing plane-parallel plates of different thickness in the plane wave illuminating the CCD sensor, figure 6.2. A plate of thickness $p$ and refraction index $n$ will produce a delay $\Delta t$ with respect to air (or the vacuum with light speed $c$) given by:

$$\Delta t = (n-1)\frac{p}{c} \tag{6.3}$$

That way it is possible to record at one time several holograms of the object using a corresponding number of reference waves delayed with respect to each other. The numerical reconstruction can be done for each part of the CCD array in which the phase of the reference wave has a particular delay, giving rise to the desired "times of evolution" of the wave front illuminating the object. This is equivalent to choose another observation point in the original LiF experiment. In this sense, the phase delays introduced in the different parts of the reference wave can be interpreted as artificial extensions of the CCD sensor and allow a better visualization of the phenomenon.

In the experiments 6 mm thick PMMA plates (refraction index n ~ 1.5) are used to produce the phase delays in the reference wave, figure 6.2. One third of the original plane reference wave does not travel through PMMA, the second third, illuminating the sensor in the middle, travels through 6 mm PMMA (representing 10*ps* delay with respect to air) and the last third travels through 18mm PMMA (30*ps* delay with respect to air). The object as seen from the CCD sensor is schematically sketched for better recognition of the results, figure 6.5. It consists of a 3cm × 3cm plane aluminum plate, which was painted matt white for better light scattering. A small plane mirror (1cm × 1cm area) is attached to the plate, perpendicular to its surface and at an angle of about 10 degrees away from the vertical.

The three reconstructed stripes of the hologram, corresponding to three different times of evolution of the wavefront illuminating the object are shown in figure 6.6. One part of the wavefront is reflected by the mirror, the other part is traveling in the original direction. The three pictures can be interpreted as a slow-motion shot of the wavefront. As demonstrated before, quantitative results can be derived from these images, e.g. the speed of light.

The minimum number of pixels required for a part of the hologram to be successfully reconstructed limits the number of different "times of evolution" that can be recorded simultaneously . Furthermore, due to the borders of the plates introduced in the reference wave diffraction effects cause dark zones at the CCD which cannot be used for numerical reconstruction.

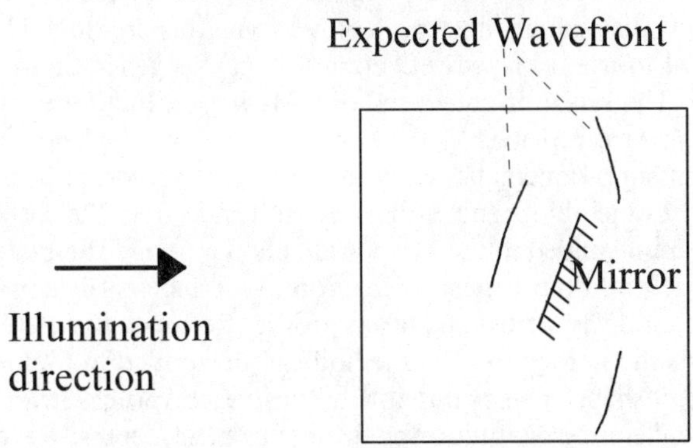


**Fig. 6.5.** Object used for displaying the temporal evolution of a wave front as seen from the CCD sensor

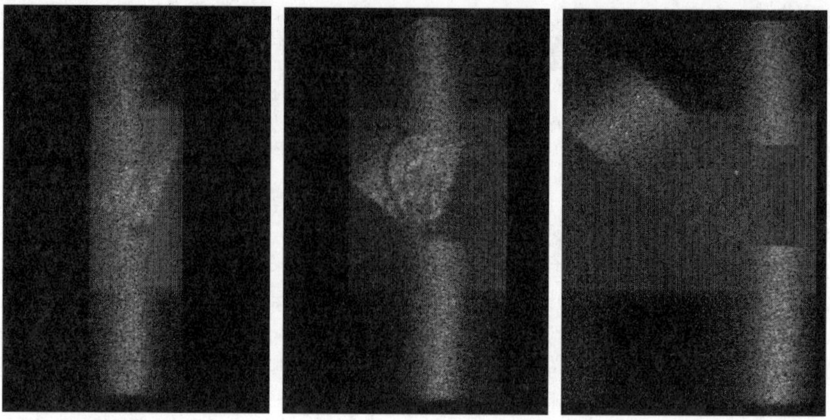

**Fig. 6.6.** The wavefront at three different times, reconstructed from one single holographic recording.
Left: No delay, wavefront just reaching mirror.
Middle: 10 ps delay, the mirror reflects one part of the wavefront.
Right: 30 ps delay with respect to the left recording, one part is reflected into the opposite direction, the other part is traveling in the original direction

### 6.1.2 Short-Coherence Tomography

The main disadvantage of introducing the path differences by glass plates with different thickness are the diffraction effects at the edges of the plates. Therefore this technique allows only a few discrete depth steps. To overcome this problem, Nilsson and Carlsson proposed to use a blazed reflection grating for generating path differences [13, 101-103]. The set-up is composed of a Michelson Interferometer, figure 6.7, in which one mirror is replaced by the grating. The incoming beam of a light source with sufficient short coherence length is split into two partial beams. One partial beam illuminates the object and is diffusely reflected from the surface to the CCD. The other beam is guided to the blazed reflection grating. The grating retro reflects that beam back into the opposite direction to the incident beam, introducing a spatially varying delay across the beam profile. Both beams interfere at the CCD, which records the hologram. The method can be applied to measure the three-dimensional object shape. This is possible because each vertical stripe of the hologram fulfils the coherence condition for different object depths. Reconstructions from different hologram parts create different depth layers of the object.

Instead of the grating in figure 6.7 it is also possible to use an ordinary mirror in the reference arm, see e. g. [117], which can be shifted in the direction of beam propagation. Digital holograms are then recorded in different mirror positions. Each single hologram represents another depth layer of the object and the shape can be calculated from the holographic reconstructions. However, there is an advantage of the setup shown in figure 6.7 using the grating: Only one recording is necessary to determine the whole shape.

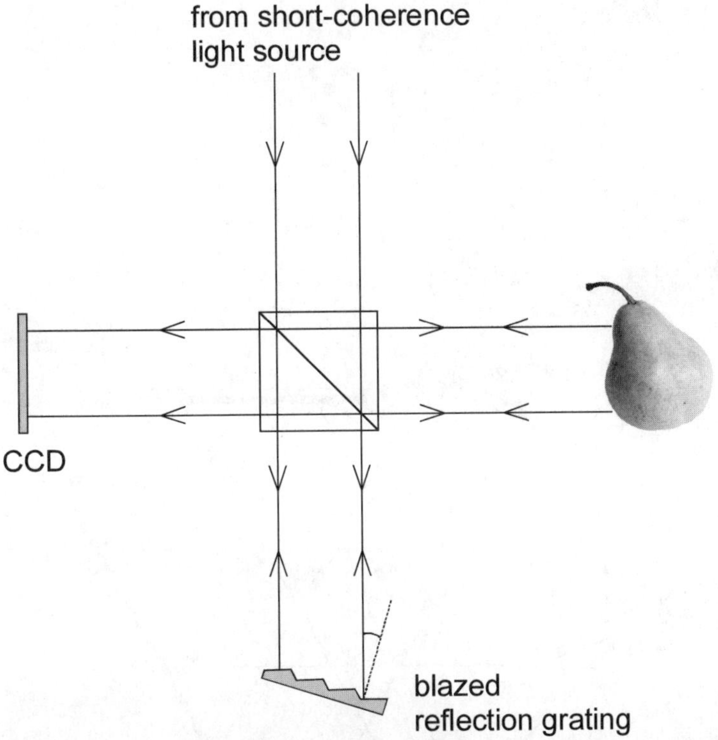


**Fig. 6.7.** Short-coherence length tomography using a blazed grating

## 6.2 Particle Distribution Measurements

The size, the speed, and the distribution of particles within fluid or gaseous media are measured with optical techniques like laser-doppler-anemometry (LDA), phase-doppler-anemometry (PDA) or particle image velocimetry (PIV). To achieve the three-dimensional spatial distribution of particles also holographic techniques are used, see corresponding articles of Trolinger [161] or Hinsch [49]. However, the development of the exposed hologram (photographic plate) and the evaluation of the distribution of the droplets are time consuming processes. Digital Holography where holograms are recorded with the CCD-sensor followed by numerical reconstruction of the object wave offers new alternatives.

Digital holograms of particle distributions within transparent media can be recorded with the simple in-line set-up depicted in figure 6.8. A plane wave illuminates the observed particles. A part of the wave is diffracted by the particles while the remaining part passes through the set-up without being diffracted. This part of the wave serves as a reference wave. The waves interfere and produce a typical diffraction pattern on a CCD-sensor.

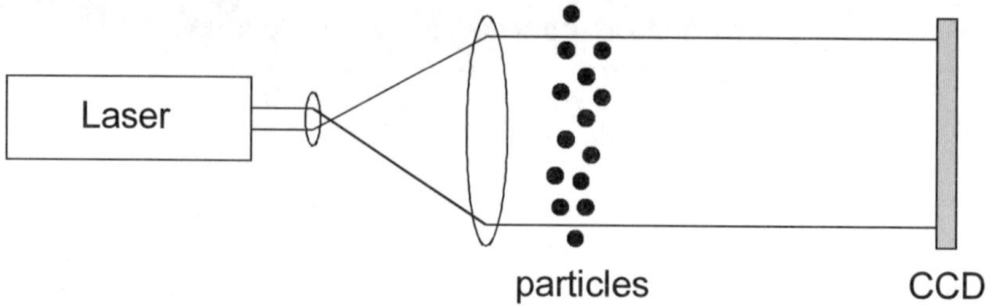


**Fig. 6.8.** In-line set-up

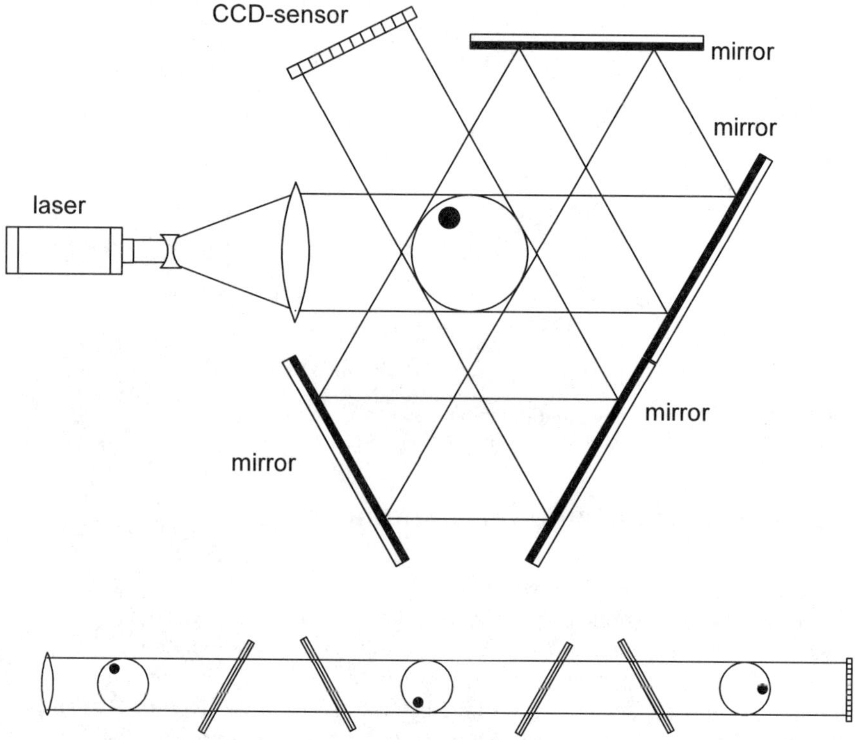


**Fig. 6.9.** Experimental set-up for recording in-line holograms from several directions with one CCD-sensor and deconvoluted lightpath (from [5])

The three-dimensional spatial distribution of small moving particles - their size and their position - can be detected by scanning the reconstruction distance in the computer: Since in numerical reconstruction the distance *d* may be chosen freely, the field can be scanned by varying *d* and check at which distances the particles are in focus.

The depth of focus in holographic imaging is a function of reconstruction distance *d*, wavelength λ and the aperture of the hologram. In Digital Holography the aperture is the CCD. In a typical example using a CCD with 18mm × 18mm area, a wavelength of λ=692nm (pulsed ruby laser) and an average reconstruction distance of *d*=0.5m the focal depth is in the range ±0.5mm. But due to the slow decrease of sharpness outside this range objects could only be detected with a positional uncertainty of about ±1cm.

In order to enhance the depth resolution, Kreis et. al. combined Digital Holography with tomographic methods [5, 6, 77]. In tomography a manifold of projections in different directions through a scene are recorded. The three-dimensional distribution of the physical quantity, e.g. the attenuation of a beam passing the scene, is then calculated by numerical methods.

To record simultaneously multiple in-line holograms of the particle stream an arrangement using a single CCD-array has been designed. After passing the particles a first time the collimated beam of a ruby pulse laser is guided by two mirrors to a second pass and by two further mirrors to a third pass through the stream before hitting the CCD-sensor. Fig. 6-9 shows schematically the dislocation of the mirrors and the deconvoluted light path. Now it is possible to extract the three views of the particles by three reconstructions with numerical focusing to the different distances. In the realized set-up the reconstruction distances are 40 cm, 65.5 cm and 95.5 cm.

The set-up is used to analyze the particle stream by Digital Holography and tomographic methods. The particles with a size of 250 μm are sprinkled manually. The diffracted light is recorded with a CCD-sensor which consists of 2048 × 2048 light-sensitive pixels. The pixel distance is 9 μm × 9 μm. The diffraction rings of each particle can be recognized in the recorded diffraction pattern (in-line hologram), figure 6.10. The more the distance of the particles to the CCD-sensor increases, the more the spacing among the concentric rings belonging to an individual particle increases as well. In addition to the diffraction rings caused by the particles, the image is overlapped with a pattern produced by the recording system itself. For protection from damage and dust, a thin glass plate covers the CCD-sensor. If illuminated with coherent light the plate produces an interference pattern. This interference pattern is even visible, if the CCD-sensor is illuminated with a plane wave and no particles are observed.

The reconstructed particles are visible as dark spots without any diffraction rings, figure 6.11. The missing diffraction rings or at least halos show that the particles are reconstructed with the appropriate distance. The lateral resolution in the reconstructed images is 9μm. The measured average particle diameter is 28 pixels, corresponding to 261μm.

The three reconstructed images show the particle stream from different directions. To gain a three-dimensional particle distribution from these images, a tomographic method is applied, Fig. 6-12.

From every image in figure 6.12 a line is taken and from the three lines a two-dimensional profile through the particle stream is calculated with a method based on the filtered backprojection approach of tomography. If this method is applied for all lines in the reconstructed images, a full three-dimensional distribution of the particles is achieved.

The three back projections of one particle must be in one plane. In the crossing of the three stripes a particle is reconstructed. The stripes arise from the low number of views used in the tomographic evaluation. With increasing number of views the stripes begin to disappear.

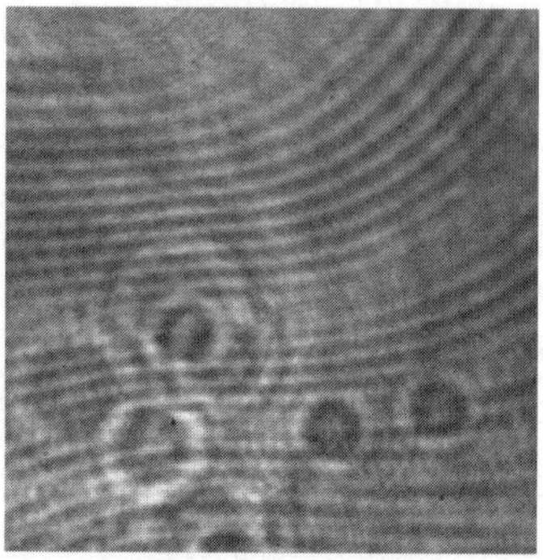

**Fig. 6.10.** In-line hologram, recorded from particles with a size of 250 µm at a distance of 40 cm, 65.5 cm and 95.5 cm

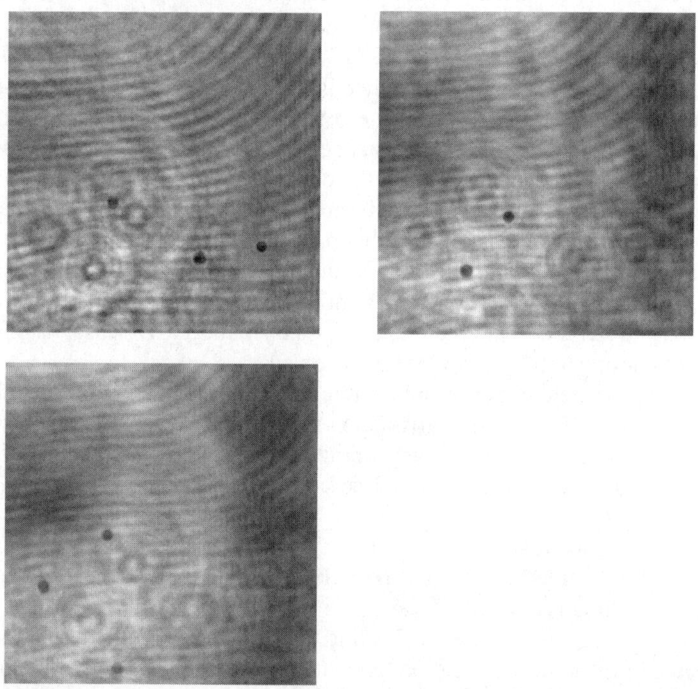

**Fig. 6.11.** Reconstructed particle distributions at a distance of 40cm, 65.5cm and 95.5cm.

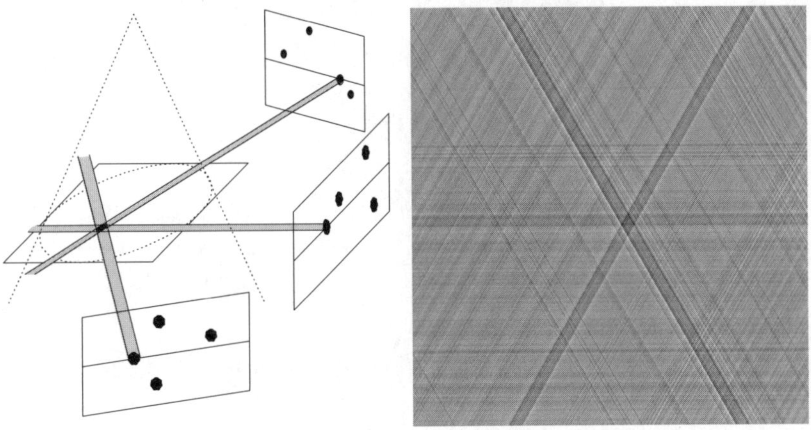

**Fig. 6.12.** *Left:* Principle of tomography. From the existing images single lines are taken and combined to a two-dimensional distribution by methods of tomography. *Right:* Two-dimensional distribution gained from three different views. In the crossing of the stripes a particle is found (from [5]).

## 6.3 Endoscopic Digital Holography

Digital Holography provides the possibility to combine deformation measurement and surface contouring in one single set-up. For measuring the object deformation in the simplest case two holograms with a certain wavelength have to be recorded for different object states (chapter 4.2). For shape measurement the object has to remain unchanged while two holograms with slightly different wavelengths or slightly different illumination points are recorded (chapter 4.3). Thus, and due to the relative simple set-up this method appears to be well suited to endoscopic measurements. It is obvious that the requirements for an endoscopic Digital Holography sensor are much higher than they are for a laboratory breadboard. Compared to a laboratory set-up the endoscopic system has to be

- more flexible;
- robust against harsh environments;
- faster in data processing;
- very small;
- adapted to restrictions caused by the system size;
- user friendly.

A sketch of a developed prototype system is shown in figure 6.13, while figure 6.14 depicts a functional prototype of the sensor head. The system can be divided in four parts: The controlling computer, the laser and the corresponding fibre coupling units, the endoscope and the sensor.

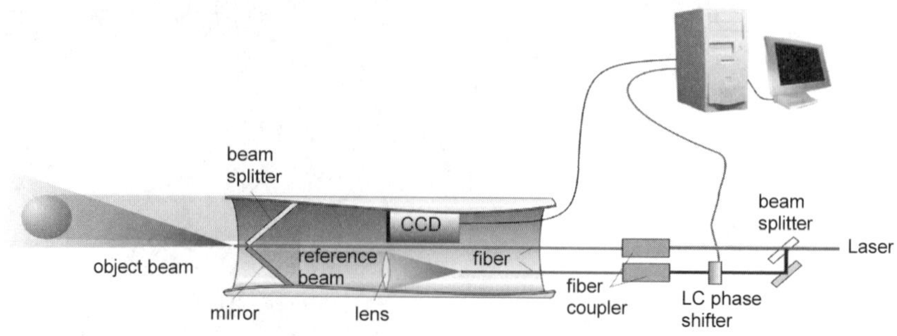


**Fig. 6.13.** Sketch of an endoscope based on Digital Holography

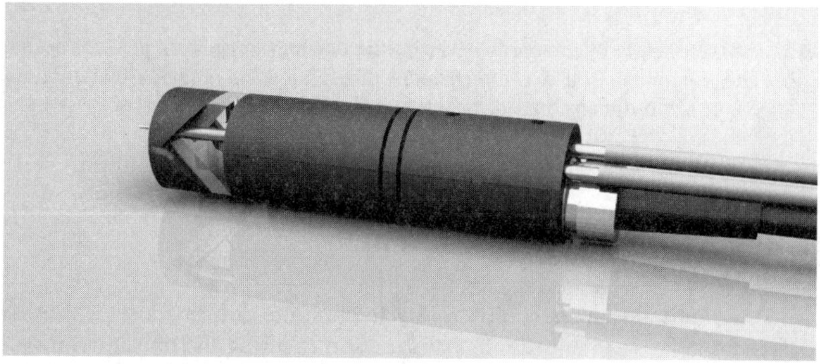

**Fig. 6.14.** Sensor head of the endoscope (photo: BIAS)

The sensor head has a diameter of 15 mm (current stage). For the future it is intended to decrease the size to a diameter of less than 10 mm.

The heart of the sensor is a commercial miniature CCD-camera with a 1/3'' CCD-matrix. Including the housing, this camera has a diameter of 10 mm. The objective of the camera is removed to be able to record the digital holograms. Since the camera provides a standard video-signal the hologram can be grabbed by a conventional frame grabber. For the object- and the reference beam mono-mode glass fibres are used. Currently, a single illumination beam is utilized. This is sufficient to measure the shape of the object and to measure one displacement component – in our case the out-of-plane component. However, in the next step three illumination directions will be implemented to be able to perform a 3D displacement measurement.

In general every laser source with sufficient coherence length can be used for deformation measurements. However, the wavelength of the laser has to be tuneable for shape measurements. Thus, to keep the whole system portable, currently a VCSEL laser diode is used as a light source. The wavelength for this laser diode type can be tuned continuously in a range of about 8 *nm*. Unfortunately, VCSEL laser diodes have a very low output power (0.5 mW @ 760 nm for single mode emission). That is why this light source can only be used for highly reflecting objects.

From figure 6.13 it can be seen that the laser beam passes through a liquid crystal phase shifter before it is coupled into the fibre for the reference beam. This LC phase shifter is used to record temporal phase shifted holograms. A simple reconstruction without using phase shifting methods results in an image that contains the desired object image and additionally the twin image together with the zero order term. By using the method of temporal phase shifting the conjugate image as well as the zero order can be eliminated completely from the resulting image (see chapter 3.3.3). In this way the entire CCD-target can be used for the measurement, i.e. the full space bandwidth of the CCD can be utilized. This is of great importance, since the choice of the camera is restricted by the system size. Cameras of this size are only available with a limited pixel number, which makes it necessary to make use of the full sensor size.

The high sensitivity of Digital Holography to object motions is also a disadvantage for a system that is intended to be used outside the laboratory. Even small object vibrations caused by environmental influences can disturb the measurement. The effect of those disturbances can be minimised by decreasing the time needed for a measurement. Thus, a high processing speed and a fast data acquisition are important to minimize the influence of unwanted vibrations. In order to achieve a high processing speed of the holograms an optimized phase-shift algorithm has been chosen [87]. More than six reconstructions per second are possible for holograms with 512 x 512 pixels using a common PC with 1.4 GHz clock frequency.

Another benefit of a high processing speed is the possibility to unwrap the phase maps of deformation measurements in real time by temporal phase unwrapping [53]. In this method the total object deformation is subdivided in many measurement steps in which the phase differences are smaller than $2\pi$. By adding up those intermediate results, the total phase change can be obtained without any further unwrapping. This is an important feature, since it is essential to have an unwrapped phase to be able to calculate the real deformation data from the phase map. Fig. 6.15 shows an example of such a measurement. The left image shows the wrapped deformation phase for a clamped coin which was loaded by heat. The right image shows the temporal unwrapped phase, which has been obtained by dividing the total deformation in 85 sub-measurements.

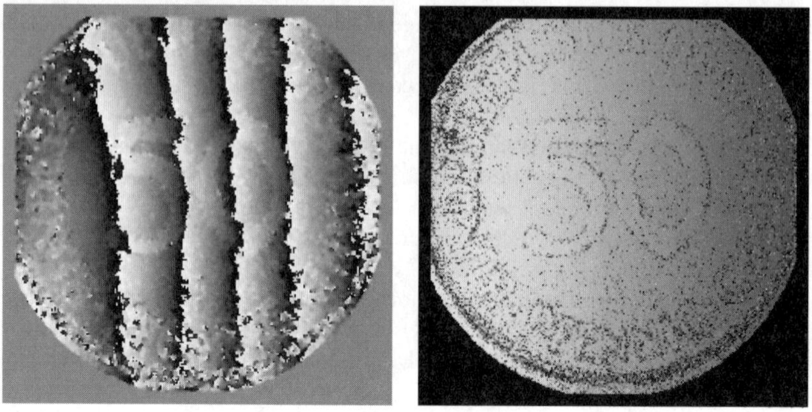

**Fig. 6.15.** Wrapped deformation phase of a heat loaded coin (left) and unwrapped phase generated by temporal unwrapping (right)

## 6.4 Optical Reconstruction of Digital Holograms

The techniques discussed in this chapter differ from the other methods described in the previous chapters, because the reconstruction is performed with an optical set-up. A computer is just used as intermediate storage medium for digital holograms.

Liquid Crystal Displays (LCD's) are electro-optical devices used to modulate light beams. Likewise they can be used as a spatial light modulator for holography. An individual LCD cell changes its transmittance depending on the applied voltage. It is therefore possible to modulate the brightness of light, which passes the device.

Optical hologram reconstruction with a LCD is possible e. g. with the set-up of figure 6.16. At first a digital hologram is recorded with a CCD-sensor, figure 6.16(a). The hologram is stored and then transmitted to the reconstruction set-up, figure 6.16(b). Recording and reconstruction set-ups could be located at different sites. The LCD modulates the reconstruction beam with the hologram function. The original object wave is reconstructed due to the diffraction of the reconstruction beam at the modulated LCD. The virtual image can be observed at the position of the original object. Usually a CCD-camera is used to record the image. Alternatively it is possible to reconstruct the real image by illuminating the LCD with the conjugate reference wave.

An example of such an optical reconstruction of a digitally recorded hologram is shown in figure 6.17(a). A digital hologram of a knight is recorded and stored. The image of the knight becomes visible if the LCD with the hologram mask is illuminated by a reconstruction wave. Optical reconstruction of two superimposed holograms, which are recorded in different object states results in a holographic interferogram, figure 6.17(b).

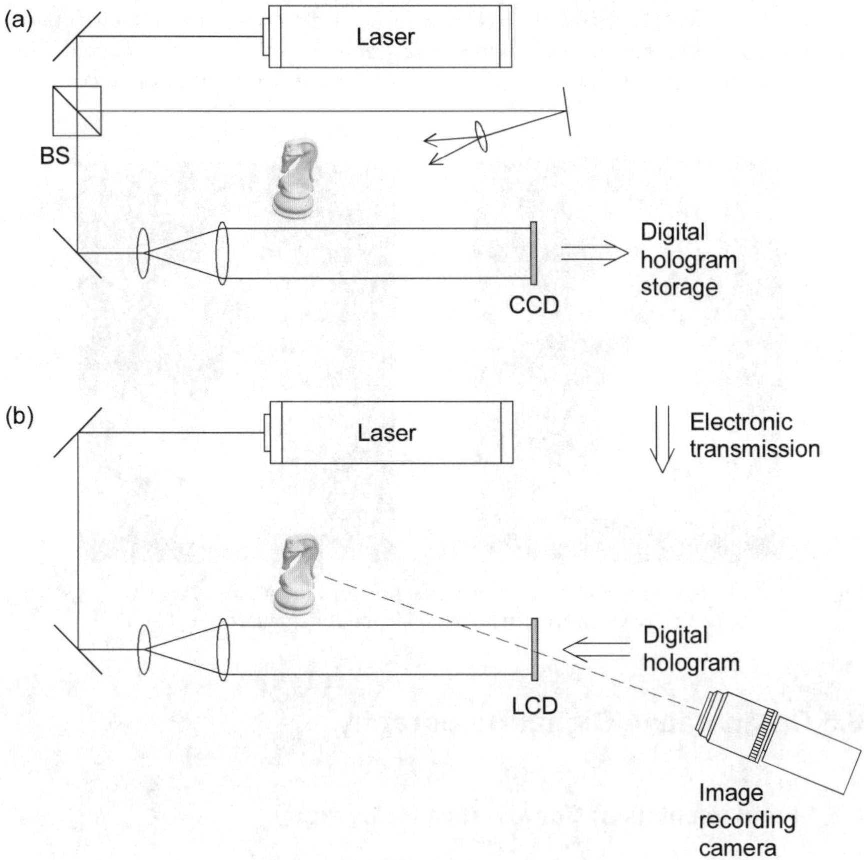


**Fig. 6.16.** (a) Digital hologram recording with a CCD
(b) Optical reconstruction with a LCD

Instead of a LCD other electro-optical devices can be used as spatial light modulators, too. Recently Kreis, Aswendt und Höfling published the optical reconstruction by means of a Digital Mirror Device (DMD) [78]. A DMD is a silicon micromachined component, which consists of an array of tiltable aluminium mirrors mounted on hinges over a CMOS static random access memory (SRAM). Today DMD's are available with up to 1280 × 1024 mirror elements. The individually addressable mirrors can be tilted binary either –10 *deg* (on) or +10 *deg* (off) along an axis diagonal to the micromirror. In an optical reconstruction set-up DMD's are therefore operated in reflection. In contrast to LCD's, which absorb up to 90% of the available light, a DMD is a reflective device yielding much more light. Consequently the diffraction efficiency in the hologram reconstruction is better compared to LCD's.

A very simple device for displaying digital holograms is a computer printer. The high resolution of standard ink-jet or laser printers with up to 3000 dots per inch makes it possible to print digital holograms directly on a transparent film. The hologram is then reconstructed by illuminating this film with the reconstruction wave.

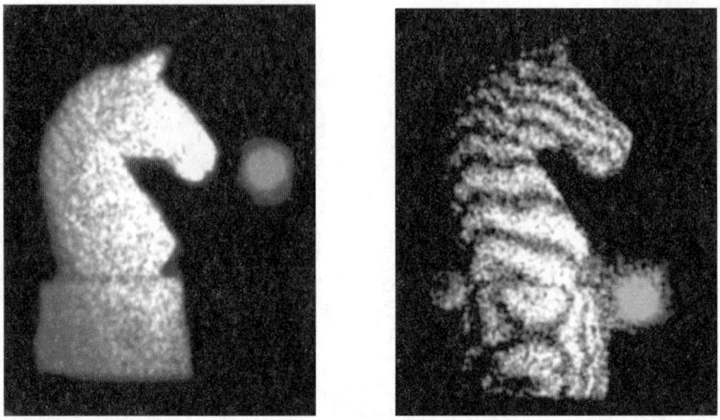

**Fig. 6.17.** (a) Optical reconstruction of a digital hologram by means of a LCD (b) Optically reconstructed holographic interferogram (from [111])

## 6.5 Comparative Digital Holography

### 6.5.1 Fundamentals of Comparative Holography

The essential of interferometry is the comparison of the optical wave reflected or transmitted by the test object with another, known wave field. In Holographic Interferometry at least one of these waves is stored by a hologram. By interference the phase differences between the two wave fields can be measured. The phase differences are connected to the quantities to be determined by the geometry function of the set-up. In this way it is possible to measure e.g. shapes or deformations of technical objects. However, a severe restriction in conventional HI is that interference is only possible, if the microstructures of the surfaces to be compared are identical. The change of the surface by the exchange of the object or e.g. too high deformations leads to a decorrelation of the two speckle fields and the loss of the interference. Thus standard HI is restricted to the comparison of two states of the *same* object.

A method to overcome this restriction is comparative interferometry [36, 100]. This method is based on the illumination of the two states of the test component with the corresponding conjugated object wave of the master object: The object wave of the master component acts as a coherent mask for the adaptive illumination of the test component.

Comparative interferometry is performed according to the following principle: At first a double exposure hologram of the master object is taken in the two states according to a specific load. The reconstruction of this double exposed hologram generates an interferogram. The relation between the measured interference phase and the displacement vector is given by Eq. (2.84):

$$\Delta\varphi_1(x,y)=\frac{2\pi}{\lambda}\vec{d}_1(x,y,z)\left(\vec{b}_1-\vec{s}_1\right)=\vec{d}_1\vec{S}_1 \tag{6.4}$$

The test object is investigated in a modified set-up for comparison with the master: It is illuminated in the original observation direction $\vec{b}_1$ by the reconstructed, conjugated wave front of the master object, i.e. the real image of the master object is projected onto the test object. It is observed in the original illumination direction $\vec{s}_1$. This results in

$$\vec{s}_2=-\vec{b}_1 \qquad \text{and} \qquad \vec{b}_2=-\vec{s}_1 \tag{6.5}$$

$$\Delta\varphi_2(x,y)=\frac{2\pi}{\lambda}\vec{d}_2(x,y,z)\left(\vec{b}_2-\vec{s}_2\right)=\frac{2\pi}{\lambda}\vec{d}_2(x,y,z)\left(\vec{b}_1-\vec{s}_1\right) \tag{6.6}$$

Since the test object is illuminated by the conjugated wave front of the master the interferogram indicates the difference of the displacements between the two objects:

$$\Delta\varphi(x,y)=\Delta\varphi_1(x,y)-\Delta\varphi_2(x,y)=\frac{2\pi}{\lambda}\left(\vec{d}_1(x,y,z)-\vec{d}_2(x,y,z)\right)\left(\vec{b}_1-\vec{s}_1\right) \tag{6.7}$$

### 6.5.2 Comparative Digital Holography

Comparative Digital Holography is a combination of comparative holography with Digital Holography [111-112]. The set-up for comparative Digital Holography follows the principle of Digital Holography and adapts the comparative method, figure 6.18. The digital hologram of the master object is recorded at a location A, figure 6.18(a). The transmission of this digital hologram to a test location B can be done by any data transfer medium, e.g. by the internet. At location B the hologram is fed into a Liquid Crystal Display as spatial light modulator. A laser reconstructs the hologram optically.

For the comparative holography the conjugated wavefronts of the master object are reconstructed and illuminate the test object, figure 6.18(b). The observation is done in the original illumination direction. A great advantage of comparative Digital Holography compared to conventional comparative HI is, that the holograms of all states can be stored and later reconstructed separately from each other. Therefore no additional reference waves are needed for the separate coding of the different holograms. This characteristic of Digital Holography reduces the technical requirements for comparative measurements significantly.

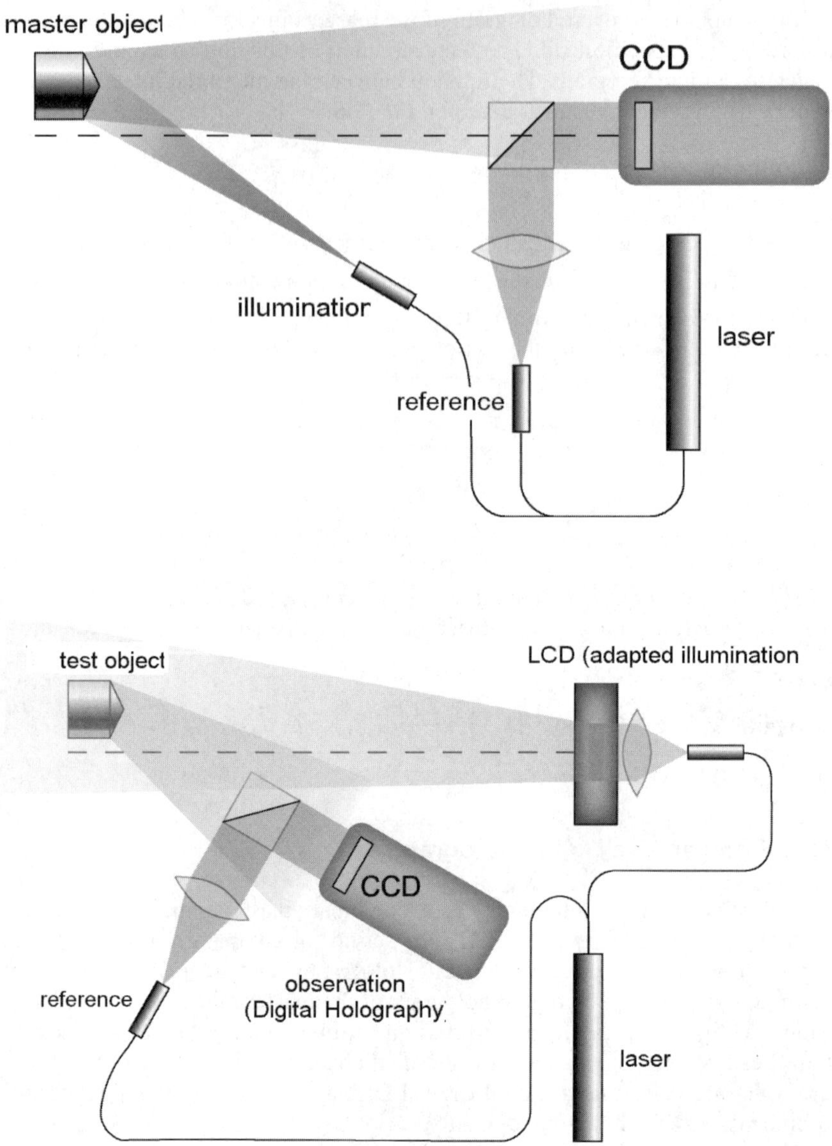


**Fig. 6.18.** Comparative Digital Holography (from [111])
(a) Recording of the mask
(b) Coherent illumination of the test object with the conjugated wavefront of the master

The method is demonstrated by the determination of a small dent in one of two macroscopically identical cylinders with cones at their upper end. The depth of the dent is about a few micrometers. With holographic two-wavelength contouring, the observed phase differences can be described by

$$\Delta\varphi_1(x,y) = \frac{2\pi}{\Lambda}\left(\vec{b}_1 - \vec{s}_1\right)\overrightarrow{\Delta r_1}(x,y) \tag{6.8}$$

$$\Delta\varphi_2(x,y) = \frac{2\pi}{\Lambda}\left(\vec{b}_2 - \vec{s}_2\right)\overrightarrow{\Delta r_2}(x,y) \tag{6.9}$$

The indices 1 and 2 mark the master or the test object, respectively. $\Lambda$ is the synthetic wavelength. The measurements are carried out with $\Lambda = 0{,}345$ mm ($\lambda_1 = 584{,}12$ nm and $\lambda_2 = 585{,}11$ nm). $\overrightarrow{\Delta r_1}$ and $\overrightarrow{\Delta r_2}$ represent the relative height deflection of the master resp. the test object. Figure 6.19(a) shows the reconstructed intensity of the test object, while the mod2π contour lines are depicted in figure 6.19(b). The dent is hardly to recognize. However, after holographic illumination of the test object with the real image of the master, the difference phase $\Delta\varphi$ corresponds to the difference in height deflections between master and test object:

$$\Delta\varphi(x,y) = \Delta\varphi_1(x,y) - \Delta\varphi_2(x,y) = \frac{2\pi}{\Lambda}\left(\overrightarrow{\Delta r_1}(x,y) - \overrightarrow{\Delta r_2}(x,y)\right)\left(\vec{b} - \vec{s}\right) \tag{6.10}$$

This phase difference is shown in figures 6.19c (mod2π-map) and 6-18d (pseudo 3D-map).

The measured phase difference distribution is quite noisy because of the large pixel sizes of the CCD target and the spatial light modulator (CCD: 9 μm, LCD: 29 μm). In the future better optical components might be available. Nevertheless, the comparison of figures 6.19b with 6-19d demonstrates the advantage of comparative Digital Holography to measure the shape difference of two objects with different microstructure: in the phase difference image the dent is clearly recognizable.

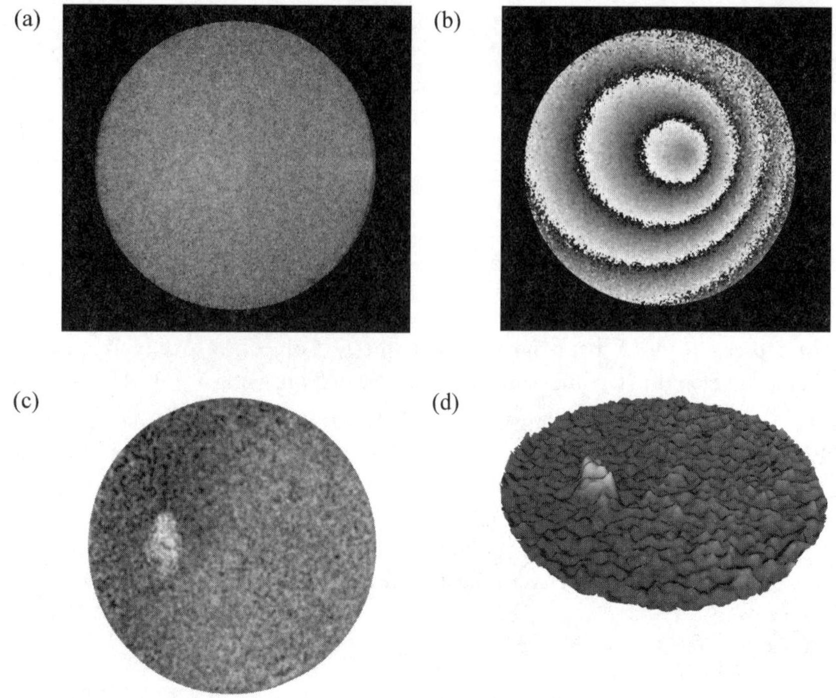

**Fig. 6.19.** Demonstration of comparative Digital Holography (from [112])
(a) Reconstructed intensity, test object
(b) Phase contour lines, test object
(c) Comparative phase difference, mod2π-map
(d) Comparative phase difference, pseudo 3D-map

## 6.6 Encrypting of Information with Digital Holography

The reconstruction of objects from their holograms is only possible, if the reconstruction wave has nearly the same properties like the reference wave used in the recording process. Any deviation from the original amplitude and phase distribution results in image aberrations or in total loss of the original object information. The reference wave used for hologram recording can be therefore regarded as a key to reconstruct the information coded in the hologram. This is the principle of information encryption by holography.

In the following a coding method proposed by Javidi et. al. [56, 155, 156] is described. The method is based on phase-shifting Digital Holography, see set-up in figure 6.20. The key for encrypting the information is a diffusely scattering screen. At first the key is recorded holographically. A parallel beam is split into two coherent partial beams at beam splitter BS1. One partial beam illuminates the screen from the back.

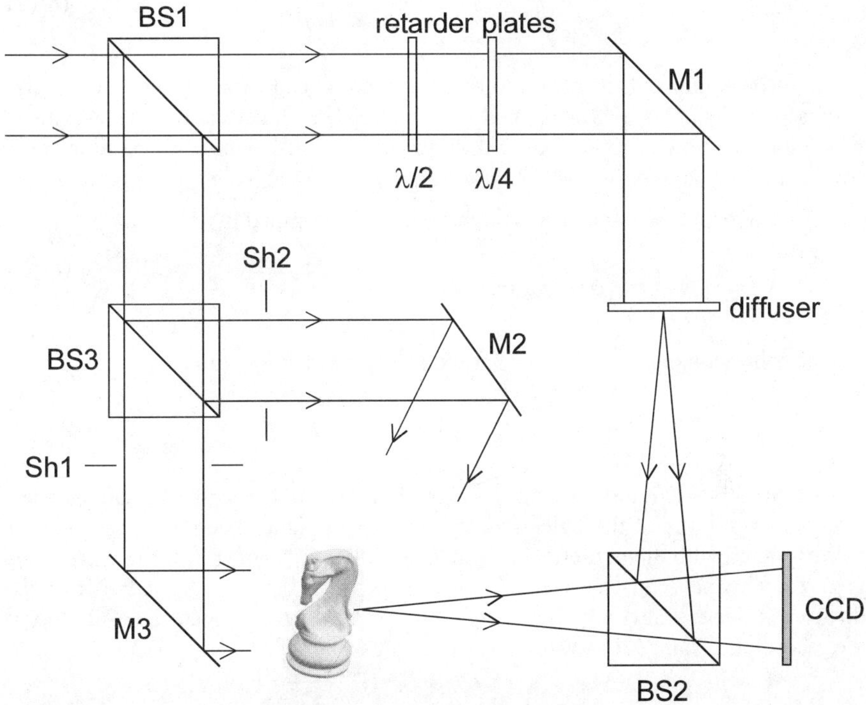


**Fig. 6.20.** Information encrypting with Digital Holography

The scattered light is guided to the CCD via beam splitter BS2. The other beam is guided via BS3, mirror M3 and BS2 to the CCD. For this shutter SH1 is opened, shutter SH2 is closed and the object is removed. Both beams interfere at the surface of the CCD. A set of four interferograms with mutual phase shifts is recorded by means of phase shifting devices. This can be done either by aligning the fast and the slow axes of optical retarders with the polarization of the incident beam (as shown in figure 6.20) or by other electro-optical devices like piezo electric driven mirrors. The complex amplitude of the plane partial wave guided via mirror M3 is $1 \cdot e^{i \cdot 0}$ in the simplest case. Consequently, it is possible to calculate the complex amplitude $a_K e^{i\varphi_K}$ of the wave scattered from the diffuser in the CCD plane by phase shifting algorithms (see also chapter 2.7.5). If four interferograms $I_1$ to $I_4$ with equidistant phase steps of $\pi/2$ are recorded the amplitude and the phase of the "key" wave are determined by following equations:

$$\varphi_K = \arctan \frac{I_4 - I_2}{I_1 - I_3} \tag{6.11}$$

$$a_K = \frac{1}{4}\sqrt{(I_1 - I_3)^2 + (I_4 - I_2)^2} \tag{6.12}$$

Now a hologram of the object to be encrypted is recorded. For this shutter SH1 is closed, and shutter SH2 is opened illuminating the object via M2. The scattered light from the screen is now used as reference wave. Again a set of four phase shifted interferograms $I_1'$ to $I_4'$ is generated. The difference phase between the "key" wave phase $\varphi_K$ and the object phase $\varphi_0$ is determined by:

$$\varphi_0 - \varphi_K = \arctan\frac{I_4'-I_2'}{I_1'-I_3'} \tag{6.13}$$

The following equation is valid for the product of the amplitudes:

$$a_0 \cdot a_K = \frac{1}{4}\sqrt{(I_1'-I_3')^2 + (I_4'-I_2')^2} \tag{6.14}$$

Without knowledge of $a_k$ and $\varphi_K$ it is obviously not possible to calculate the complex amplitude of the object wave in the CCD plane. The object can only be reconstructed with the numerical methods described in chapter 3, if the correct key is given. This key consists of amplitude and phase of the wave scattered from the diffuser. A second key, which has to be known for correct object decoding, too, is the recording distance $d$ between diffuser and CCD.

## 6.7 Synthetic Apertures

Every part of a hologram contains the entire object information. The object can be reconstructed therefore from any cut-out of the hologram, the size of such a cut-out only influences the speckle size in the reconstructed image. On the other hand it is also possible to synthesize a hologram from different single holograms [79]. A possible recording geometry with two CCD's is depicted in figure 6.21 for a plane reference wave. Independent CCD-cameras are used to record the single holograms. A fixed phase relation between the individual holograms is ensured by using the same reference wave. Reconstruction of the synthesized hologram is possible by following methods: The single holograms are reconstructed separately and the resulting complex amplitudes in the image plane are coherently superimposed. A second possibility is to embed both single holograms in an artificial large hologram, where the grey values of all pixels not covered are set to zero (black). Such artificial hologram matrix is then reconstructed as a whole.

The resolution of images and phase maps reconstructed from digital holograms depends on the recording distance $d$ and on the effective aperture $N\Delta x$, see Eq. (3.23). However, both quantities are not independent from each other, because for state of the art CCD's with pixel sizes in the range of $5\mu m$ large apertures require also long recording distances due to the spatial frequency limitations discussed in chapter 3.4.2. Increasing the aperture size by using more than one CCD therefore does not automatically improve the image resolution, because the larger synthetic aperture requires a longer recording distance. In order to decouple recording distance and aperture size it is therefore necessary to use CCD's with small pixel sizes in the range of one micron or below, which might be available in the future. With such devices even the highest spatial frequencies could be resolved, independently from the recording distance.

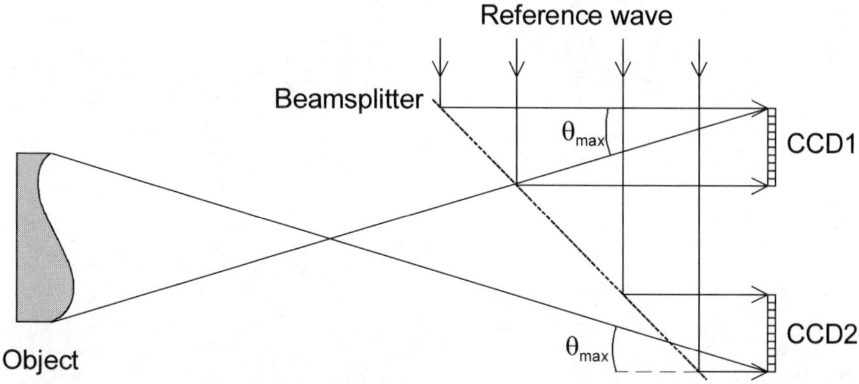


**Fig. 6.21.** Aperture synthesis with two CCD's

# 7 Speckle Metrology

## 7.1 Electronic Speckle Pattern Interferometry (ESPI)

Electronic Speckle Pattern Interferometry (ESPI) is a method, similar HI, to measure optical path changes caused by deformation of opaque bodies or refractive index variations within transparent media [48, 94]. In ESPI electronic devices (CCD's) are used to record the information. The speckle patterns which are recorded by an ESPI system can be considered as image plane holograms. Image plane holograms are holograms of focussed images. Due to the digital recording and processing, ESPI is designated also as *Digital Speckle Pattern Interferometry* (DSPI). Another designation is *TV-holography*. However, instead of hologram reconstruction the speckle pattern are correlated.

The principal set-up of an Electronic Speckle Pattern Interferometer is shown in figure 7.1. The object is imaged onto a camera (CCD) by a lens system. Due to the coherent illumination the image is a speckle pattern. According to Eq. (2.57) the speckle size depends on the wavelength, the image distance and the aperture diameter. The speckle size should match with the resolution (pixel size) of the electronic target. This can be achieved by closing the aperture of the imaging system.

The speckle pattern of the object surface is superimposed on the target with a spherical reference wave. The source point of the reference wave should be located in the centre of the imaging lens. Due to this in-line configuration the spatial frequencies are resolvable by the CCD. In practice the reference wave is coupled into the set-up by a beam splitter (as shown in figure 7.1) or guided via an optical fibre, which is mounted directly in the aperture of the lens system.

The intensity on the target is:

$$\begin{aligned} I_A(x,y) &= \left|a_R(x,y)\exp(i\varphi_R) + a_O(x,y)\exp(i\varphi_O)\right|^2 \\ &= a_R^{\ 2} + a_O^{\ 2} + 2a_R a_O \cos(\varphi_O - \varphi_R) \end{aligned} \tag{7.1}$$

$a_R \exp(i\varphi_R)$ is the complex amplitude of the reference wave and $a_O \exp(i\varphi_O)$ is the complex amplitude of the object wave in the image plane. The term $(\varphi_O - \varphi_R)$ is the phase difference between reference and object wave, which varies randomly from point to point. This speckle interferogram is recorded and electronically stored.

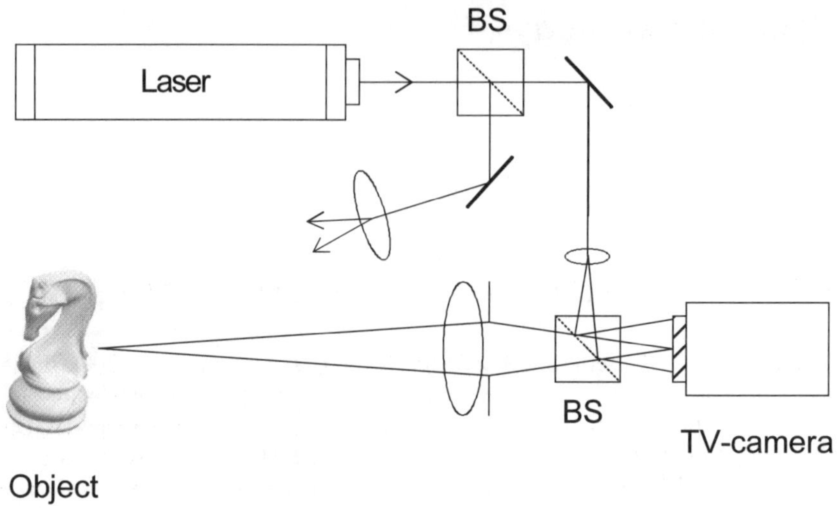


**Fig. 7.1.** Electronic Speckle Pattern Interferometer

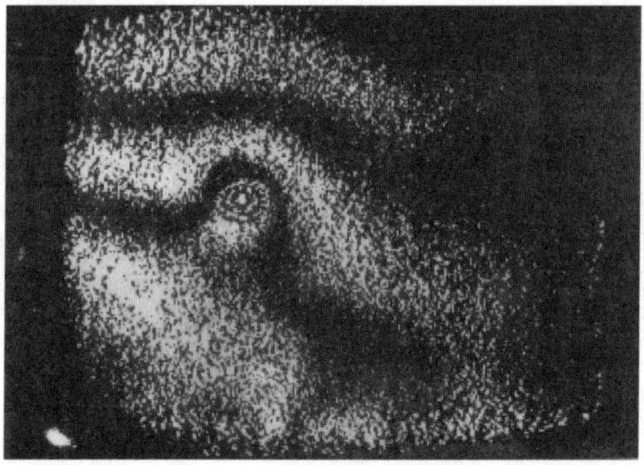

**Fig. 7.2.** ESPI image

The set-up in figure 7.1 is sensitive to out-of-plane deformations, i. e. deformations perpendicular to the object surface. A displacement of $d_z$ corresponds to a phase shift of

$$\Delta\varphi = \frac{4\pi}{\lambda} d_z \tag{7.2}$$

After deformation a second speckle pattern is recorded:

$$I_B(x,y) = \left|a_R(x,y)\exp(i\varphi_R) + a_O(x,y)\exp(i\varphi_O + \Delta\varphi)\right|^2 \\ = a_R^{\ 2} + a_O^{\ 2} + 2a_R a_O \cos(\varphi_O - \varphi_R + \Delta\varphi) \tag{7.3}$$

These two speckle pattern are now subtracted:

$$\Delta I = \left|I_A - I_B\right| = \left|2a_R a_O\left(\cos(\varphi_O - \varphi_R) - \cos(\varphi_O - \varphi_R + \Delta\varphi)\right)\right| \\ = 2a_R a_O\left|\sin\left(\varphi_0 - \varphi_R + \frac{\Delta\varphi}{2}\right)\sin\frac{\Delta\varphi}{2}\right| \tag{7.4}$$

The intensity of this difference image is minimal at those positions, where $\Delta\varphi = 0, 2\pi, \ldots$. The intensity reaches its maximum at those positions, where $\Delta\varphi = \pi, 3\pi, \ldots$. The result is a pattern of dark and bright fringes, similar to a holographic interferogram. However, differences to HI are the speckle appearance of the fringes and the loss of the three-dimensional information in the correlation process. A typical ESPI subtraction pattern is shown in figure 7.2.

As already mentioned, the set-up of figure 7.1 is only sensitive to out-of plane motions. In-plane displacements can be measured with the set-up of figure 7.3. Two plane waves illuminate the object symmetrically at the angles $\pm\theta$ to the z-axis. The object is imaged by a camera. Again the speckle size is adapted to the target resolution by the aperture of the imaging system. The phase change due to an in-plane displacement can be derived by geometrical considerations, similar to the HI displacement calculations. The phase change of the upper beam is

$$\Delta\varphi_1 = \frac{2\pi}{\lambda}\vec{d}\left(\vec{b} - \vec{s}_1\right) \tag{7.5}$$

with displacement vector $\vec{d}$. The unit vectors $\vec{b}$, $\vec{s}_1$ and $\vec{s}_2$ are defined in figure 7.3. The corresponding phase shift of the lower beam is

$$\Delta\varphi_2 = \frac{2\pi}{\lambda}\vec{d}\left(\vec{b} - \vec{s}_2\right) \tag{7.6}$$

The total phase shift is

$$\Delta\varphi = \Delta\varphi_1 - \Delta\varphi_1 = \frac{2\pi}{\lambda}\vec{d}\left(\vec{s}_2 - \vec{s}_1\right) \tag{7.7}$$

The vector $(\vec{s}_2 - \vec{s}_1)$ is parallel to the x-axis, its length is $2\sin\theta$. The result for the total phase shift as measured by the camera is therefore:

$$\Delta\varphi = \frac{4\pi}{\lambda}d_x \sin\theta \tag{7.8}$$

By using non-symmetrical illumination, the method also becomes sensitive to out-of-plane displacements.

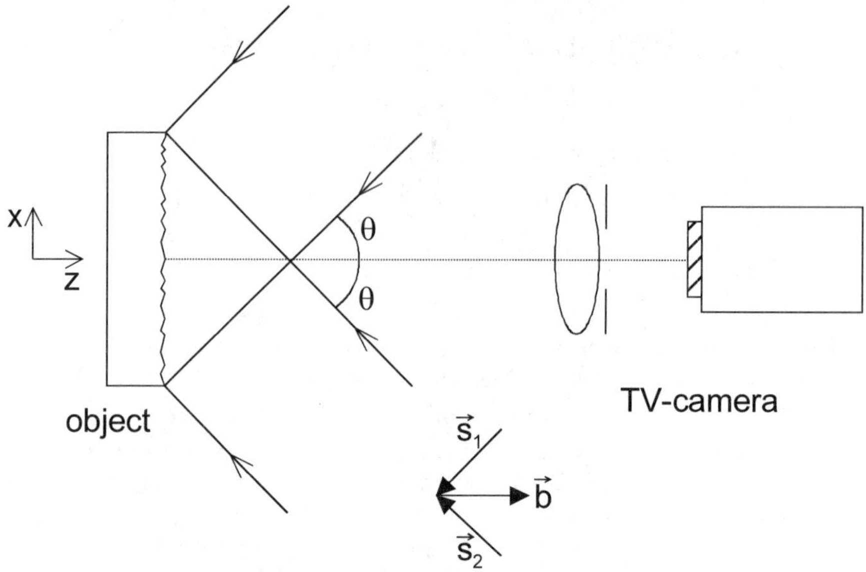


**Fig. 7.3** In-plane sensitive speckle interferometer

As for HI the phase cannot be determined from a single speckle pattern. The interference phase has to be recovered e. g. with phase shifting methods [19, 152, 153]. Phase shifting ESPI requires to record at least 3 speckle interferograms with mutual phase shifts *in each state*. Any of the various phase shifting methods can be applied. Here the algorithm with 4 recordings and unknown, but constant phase shift angle α is used. The equation system for the initial state is:

$$\begin{aligned} I_{A,1} &= a_R^2 + a_O^2 + 2a_R a_O \cos(\varphi_0 - \varphi_R) \\ I_{A,2} &= a_R^2 + a_O^2 + 2a_R a_O \cos(\varphi_0 - \varphi_R + \alpha) \\ I_{A,3} &= a_R^2 + a_O^2 + 2a_R a_O \cos(\varphi_0 - \varphi_R + 2\alpha) \\ I_{A,4} &= a_R^2 + a_O^2 + 2a_R a_O \cos(\varphi_0 - \varphi_R + 3\alpha) \end{aligned} \tag{7.9}$$

The dependence of the intensities and amplitudes from the spatial coordinates *(x,y)* has been omitted. This equation system has following solution:

$$\varphi_O - \varphi_R = \arctan \frac{\sqrt{I_{A,1} + I_{A,2} - I_{A,3} - I_{A,4}} \cdot \sqrt{3I_{A,2} - 3I_{A,3} - I_{A,1} + I_{A,4}}}{I_{A,2} + I_{A,3} - I_{A,1} - I_{A,4}} \tag{7.10}$$

For the second state also 4 phase shifted interferograms are recorded:

$$I_{B,1} = a_R^2 + a_O^2 + 2a_R a_O \cos(\varphi_0 - \varphi_R + \Delta\varphi) \tag{7.11}$$
$$I_{B,2} = a_R^2 + a_O^2 + 2a_R a_O \cos(\varphi_0 - \varphi_R + \Delta\varphi + \alpha)$$
$$I_{B,3} = a_R^2 + a_O^2 + 2a_R a_O \cos(\varphi_0 - \varphi_R + \Delta\varphi + 2\alpha)$$
$$I_{B,4} = a_R^2 + a_O^2 + 2a_R a_O \cos(\varphi_0 - \varphi_R + \Delta\varphi + 3\alpha)$$

The solution is:

$$\varphi_O - \varphi_R + \Delta\varphi = \arctan \frac{\sqrt{I_{B,1} + I_{B,2} - I_{B,3} - I_{B,4}} \cdot \sqrt{3I_{B,2} - 3I_{B,3} - I_{B,1} + I_{B,4}}}{I_{B,2} + I_{B,3} - I_{B,1} - I_{b,4}} \tag{7.12}$$

The interference phase $\Delta\varphi$ is now calculated from Eq. (7.10) and (7.12) by subtraction.

Phase shifting speckle interferometry is sometimes also called *Electro-Optic Holography* (EOH).

## 7.2 Digital Shearography

ESPI as well as conventional and Digital HI are very high sensitive to optical path changes. Displacement measurements with a resolution of up to one hundredth of the wavelength are possible. On the other hand this high sensitivity is also a drawback for applications in a rough environment, where no vibration isolation is available. Unwanted optical path length variations due to vibrations disturb the recording process.

*Shearography* [8, 51, 52, 82, 108] is an interferometric method, which brings the rays scattered from one point of the object $P(x, y)$ into interference with those from a neighbouring point $P(x + \Delta x, y)$. The distance between both points is $\Delta x$. The shearing can be realized e. g. by a glass wedge, mounted in one half of the imaging system, figure 7.4. The object is imaged via both halfes of the aperture (with and without wedge). Therefore two laterally sheared images overlap at the recording device, see figure 7.5.

The intensity on the target is:

$$I_A(x, y) = |a_1(x, y)\exp(i\varphi(x, y)) + a_2(x, y)\exp(i\varphi(x + \Delta x, y))|^2 \tag{7.13}$$
$$= a_1^2 + a_2^2 + 2a_1 a_2 \cos(\varphi(x, y) - \varphi(x + \Delta x, y))$$

where $a_1 \exp(i\varphi(x, y))$ and $a_2 \exp(i\varphi(x + \Delta x, y))$ are the complex amplitudes of the interfering waves in the image plane. As for ESPI the phase difference $(\varphi(x + \Delta x, y) - \varphi(x, y))$ varies randomly from point to point. This speckle interferogram is recorded and electronically stored. Another interferogram is recorded for the second state B:

$$
\begin{aligned}
&I_B(x,y) \qquad (7.14)\\
&= \begin{vmatrix} a_1(x,y)\exp[i(\varphi(x,y)+\Delta\varphi(x,y))] \\ +\, a_2(x,y)\exp[i(\varphi(x+\Delta x,y)+\Delta\varphi(x+\Delta x,y))] \end{vmatrix}^2 \\
&= a_1{}^2 + a_2{}^2 \\
&+ 2a_1a_2\cos[\varphi(x,y)-\varphi(x+\Delta x,y)+\Delta\varphi(x,y)-\Delta\varphi(x+\Delta x,y)]
\end{aligned}
$$

Pointwise subtraction gives:

$$
\begin{aligned}
&\Delta I = |I_A - I_B| \qquad (7.15)\\
&= \left| 2a_1a_2 \left\{ \cos(\varphi(x,y)-\varphi(x+\Delta x,y)) - \cos\begin{bmatrix} \varphi(x,y)-\varphi(x+\Delta x,y) \\ +\Delta\varphi(x,y)-\Delta\varphi(x+\Delta x,y) \end{bmatrix} \right\} \right| \\
&= 2a_1a_2 \begin{vmatrix} \sin\left\{ \varphi(x,y)-\varphi(x+\Delta x,y)+\dfrac{\Delta\varphi(x,y)-\Delta\varphi(x+\Delta x,y)}{2} \right\} \\ \times \sin\dfrac{\Delta\varphi(x,y)-\Delta\varphi(x+\Delta x,y)}{2} \end{vmatrix}
\end{aligned}
$$

This correlation pattern is named *shearogram*, see typical example in figure 7.6.

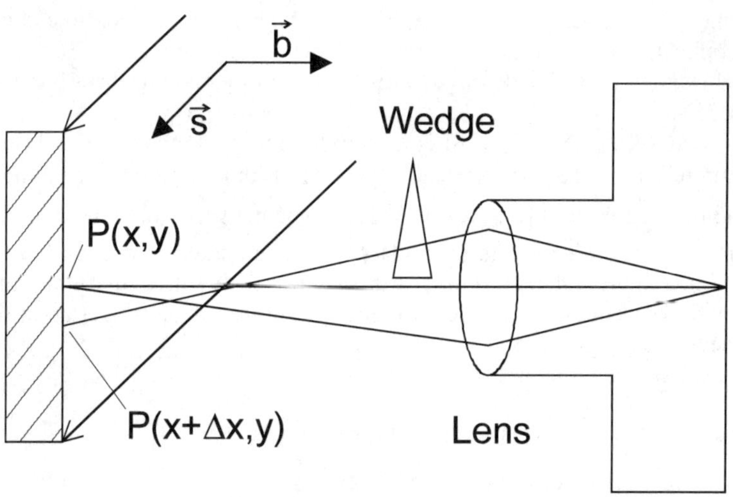


**Fig. 7.4.** Speckle shearing interferometer

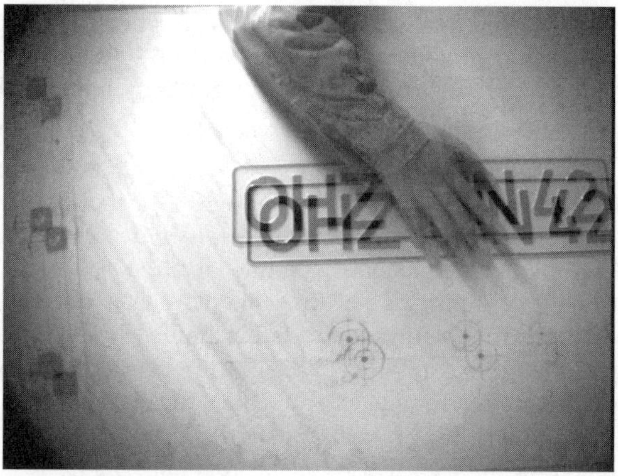

**Fig. 7.5.** Image of a shearing camera

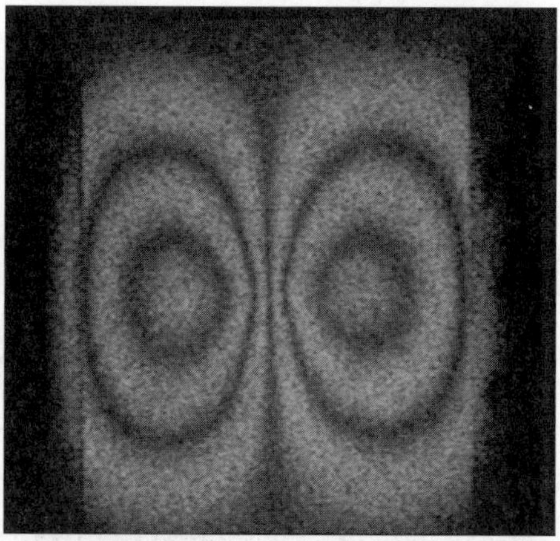

**Fig. 7.6.** Shearogram

The phase shift due to deformation in the argument of Eq. (7.15) is calculated as follows (see also the definition of unit vectors $\vec{b}$ and $\vec{s}$ in figure 7.4):

$$\begin{aligned}
&\Delta\varphi(x,y)-\Delta\varphi(x+\Delta x,y) \\
&=\frac{2\pi}{\lambda}\left\{\vec{d}(x,y)\left(\vec{b}-\vec{s}\right)-\vec{d}(x+\Delta x,y)\left(\vec{b}-\vec{s}\right)\right\} \\
&=\frac{2\pi}{\lambda}\left\{\frac{\vec{d}(x,y)-\vec{d}(x+\Delta x,y)}{\Delta x}\left(\vec{b}-\vec{s}\right)\right\}\Delta x \\
&\approx\frac{2\pi}{\lambda}\frac{\partial\vec{d}(x,y)}{\partial x}\left(\vec{b}-\vec{s}\right)\Delta x
\end{aligned} \tag{7.16}$$

A shearing interferometer is therefore sensitive to the derivative of the displacement into the shear direction, in contrast to ESPI which is sensitive to the displacement. Shearography is relatively insensitive for rigid body motions, because $\partial \vec{d}(x,y)/\partial x$ vanishes if the object is moved as a whole [149, 150]. A second property which makes a shearing interferometer less sensitive to vibrations is the self-reference principle: Optical path changes due to vibrations influence both partial beams, which means they compensate each other to a certain degree. Shearography is therefore suited for rough environments with low vibration isolation.

The measurement sensitivity of a speckle shearing interferometer can be adjusted by varying the shearing $\Delta x$. This parameter is determined by the wedge angle in the interferometer set-up of figure 7.4. Other shearing interferometer geometries are based on a Michelson interferometer, where the mirror tilt determines the shearing, figure 7.7.

The phase shifting techniques can be used also for shearography. As for ESPI a set of phased shifted images is recorded in each state from which the phase according to Eq. (7.16) is calculated.

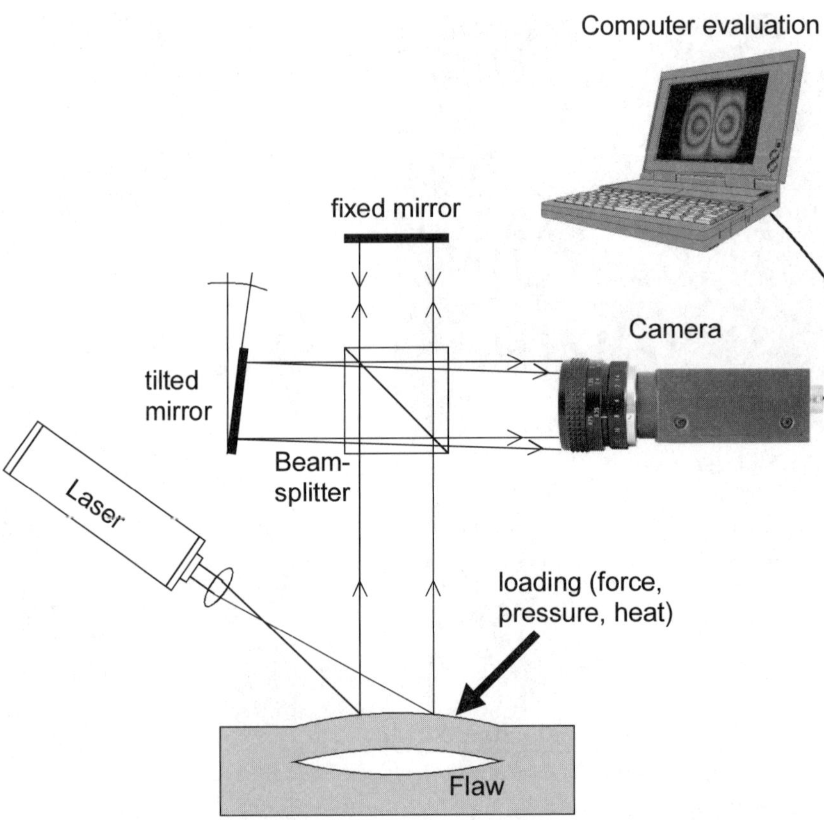


**Fig. 7.7.** Shearography set-up based on a Michelson interferometer

## 7.3 Digital Speckle Photography

Digital Speckle Photography (DSP) is the electronic version of Speckle Photography [15, 50, 83, 145, 154]. The method is used to measure in-plane displacements and strains. In classical Speckle Photography two speckle patterns of the same surface are recorded on photographic film, e. g. with the set-up of figure 2.9. The object suffers an in-plane deformation between the exposures. This in-plane deformation is made visible as fringe pattern by pointwise illumination of the double exposed film with a collimated laser beam or alternatively using an optical filtering set-up. In DSP the speckle patterns are recorded by a high resolution CCD-camera, electronically stored and correlated numerically. DSP has the potential to measure under dynamic testing conditions, because a single recording at each load state is sufficient for the evaluation. Furthermore, the requirements for vibration isolation are much lower than for interferometric methods, because DSP works without reference wave. DSP is therefore an attractive tool for measurements under workshop conditions.

The sample under investigation is coherently illuminated by means of an expanded laser beam. A speckle pattern of the reference state and a speckle pattern of the load state are recorded. The first step of the numerical evaluation procedure is to divide the whole image of e. g. 2024 × 2024 pixels into subimages, figure 7.8. Usual sizes of these subimages are 64 × 64 pixels or 32 × 32 pixels. The calculation of the local displacement vectors at each subimage is performed by a cross correlation function

$$R_{II}(d_x, d_y) = \int_{-\infty}^{\infty}\int_{-\infty}^{\infty} I_1^*(x,y) I_2(x+d_x, y+d_y)\, dxdy \tag{7.17}$$

where $I_1(x,y)$ and $I_2(x,y)$ are the intensities in the reference and in the load speckle pattern, respectively. The quantities $d_x$ and $d_y$ are the displacements of the subimage in *x*- and *y*-direction. Intensities are always real, i. e. the conjugate complex operation can be neglected. A mathematical equivalent form of Eq. (7.17) is:

$$R_{II}(d_x, d_y) = \Im^{-1}\left\{\Im\left[I_1^*(x,y)\right]\Im\left[I_2(x,y)\right]\right\} \tag{7.18}$$

The mean displacement vector of the evaluated subimage is given by the location of the peak of the cross correlation function, figure 7.8. This numerical evaluation corresponds to the classical technique, where double exposed speckle photos are locally illuminated by a collimated laser beam. The full in-plane displacement map of the monitored area is available after evaluation of all subimages.

The displacement field is calculated by this methods in integer numbers of one pixel. The accuracy is therefore only in the order of one pixel. This discrete evaluation is sufficient for applications where only displacements fields are to be measured. Strain analyses of experimental mechanics require often a higher measurement accuracy, because differences have to be calculated. The normal strains are e. g. given by

$$\varepsilon_x = \frac{\partial d_x}{\partial x} \approx \frac{\Delta d_x}{\Delta x} \quad ; \quad \varepsilon_y = \frac{\partial d_y}{\partial y} \approx \frac{\Delta d_y}{\Delta y} \tag{7.19}$$

The accuracy of DSP can be improved using so called subpixel algorithms, where the displacements are calculated on the basis of all floating point values in the neighborhood of the pixel with the peak location. A simple subpixel algorithm is given by

$$d_x = \frac{\sum_i d_{x,i} G_i}{\sum_i G_i} \quad d_y = \frac{\sum_i d_{y,i} G_i}{\sum_i G_i} \tag{7.20}$$

where $G_i$ is the floating point grey level of pixel number $i$. The structure of this formula is equivalent to a "center of gravity" calculation. In practice only a few pixels around the peak are necessary for the subpixel evaluation.

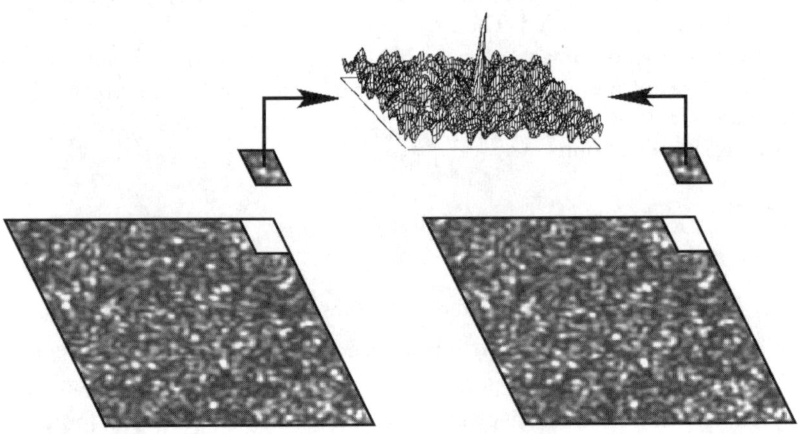

**Fig. 7.8.** Cross correlation of subimages

## 7.4 Comparison of Conventional HI, ESPI and Digital HI

Conventional Holographic Interferometry using photographic or other recording media, Electronic Speckle Pattern Interferometry and Digital Holographic Interferometry are different methods to measure optical path changes. In this chapter the differences as well as the common features of all three methods are analyzed.

The process flow diagram of conventional real-time HI is shown in figure 7.9. The measurement process starts by recording of a hologram of the object in its initial state. This is the most time-consuming and costly step of the entire process:

The exposure, development and repositioning of a photographic plate takes typically some minutes. Other holographic recording media such as thermoplastic films or photorefractive crystals can be developed much faster and automated (some seconds for thermoplastic films, even instantaneously for photorefractive crystals). However, the quality and reliability of thermoplastic films is not sufficient for HI applications and the information stored in photorefractive crystals erases in the optical reconstruction process. Once the hologram is successfully developed and replaced at its initial position the process is simple: The superposition of the wave field reconstructed by the hologram with the actual wave field generates a holographic interferogram. This fringe pattern is recorded by an electronic camera and digitally stored. In order to determine the interference phase unambiguously, it is necessary to generate at least three interferograms (for the same object state) with mutual phase shifts. The interference phase is calculated from these phase shifted interferograms by the algorithm briefly discussed in chapter 2.7.5. The entire process requires altogether the generation of one hologram plus recording of at least three interferograms in order to determine the interference phase. The technical effort is tremendous: A holographic set-up with interferometer and laser, holographic recording media (photographic plates), laboratory equipment for development of holograms, a phase shifting unit and an electronic camera with storage device (PC) for interferogram recording are necessary. On the other hand the quality of interferograms generated by this method is excellent. Due to the size and resolution of holograms recorded on photographic plates the observer can choose the observation direction freely, i. e. it is possible to observe the object from a variety of different positions and with different depth of focus. This is often very helpful for NDT applications and, if the sensitivity vector has to be varied, in quantitative deformation measurement.

ESPI was born from the desire to replace photographic hologram recording and processing by recording with electronic cameras. At the beginning of the seventieth of the last century, when ESPI was invented, only analogous cameras with very low resolution (linepairs per millimetre) were available. Consequently, a direct conversion of holographic principles to electronic recording devices was not possible. The basic idea of ESPI therefore was to record holograms of focussed images. The spatial frequencies of these image plane holograms could be adapted to the resolution of the cameras due to the in-line configuration. The optical reconstruction was replaced by an image correlation (subtraction). The ESPI correlation patterns have some similarities to the fringes of HI, but have a speckle appearance. Another difference to conventional HI is the loss of the 3D-effect, because only image plane holograms are recorded from one observation direction. Interference phase measurement with ESPI require application of phase shifting methods, see process flow in figure 7.10. In each state at least three speckle interferograms with mutual phase shifts have to be recorded. The total number of electronic recordings to determine the interference phase is therefore at least six. Speckle interferometers are commercially available. These devices can be used nearly as simple as ordinary cameras.

The idea of Digital Holographic Interferometry was to record "real" holograms (not holograms of focussed images) by an electronic device and to transfer the optical reconstruction process into the computer. The method is characterized by following features:

- No wet-chemical or other processing of holograms (as for ESPI)
- From one digital hologram different object planes can be reconstructed by numerical methods (numerical focussing)
- Lensless imaging, i. e. no aberrations by imaging devices
- Direct phase reconstruction, i. e. phase differences can be calculated directly from holograms, without interferogram generation and processing. This interesting feature is only possible in DHI, conventional HI as well as ESPI need phase shifted interferograms (or another additional information) for phase determination. The total number of recordings to get the interference phase is therefore only two (one per state), see process flow in figure 7.11. Even transient processes, where there is no time for recording of phase shifted interferograms, can be investigated with DHI.

DHI and phase shifting ESPI are competing techniques. ESPI is working since many years in real-time, i. e. the recording speed is only limited by the frame rate of the recording device (CCD). In addition the user sees directly an image of the object under investigation, while this image is only available in DHI after running the reconstruction algorithm. This *what you see is what you get* feature is helpful for adjustment and control purposes. On the other hand the time for running the DHI reconstruction algorithms has been reduced drastically in recent years due to the progress in computer technology. Digital holograms with 1000 × 1000 pixels can nowadays be reconstructed also nearly in real-time.

Another slight *present* disadvantage of DHI is that the spatial frequency spectrum has to be adapted carefully to the resolution (pixel size) of the CCD. However, also the CCD technique makes progress and CCD's with further decreasing pixel distances may be expected in future, allowing a greater object angle cone.

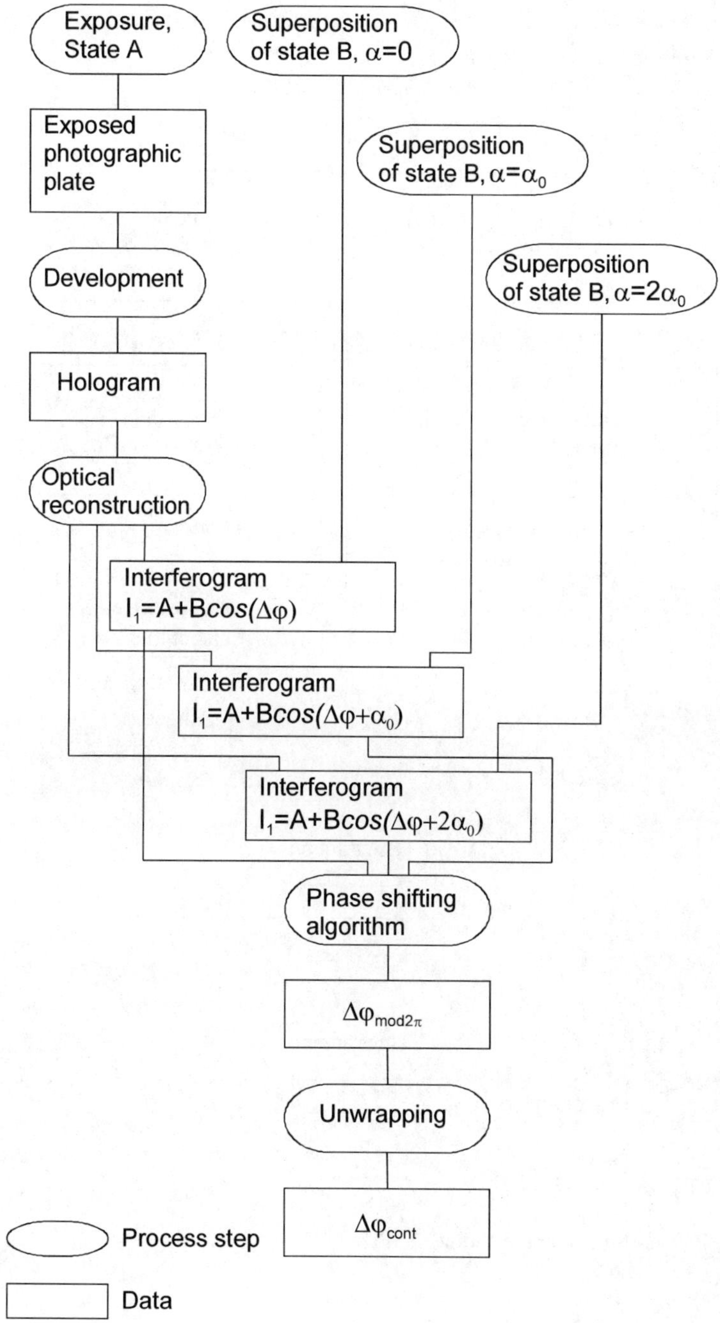


**Fig. 7.9.** Process flow of real-time HI with phase shifting fringe evaluation

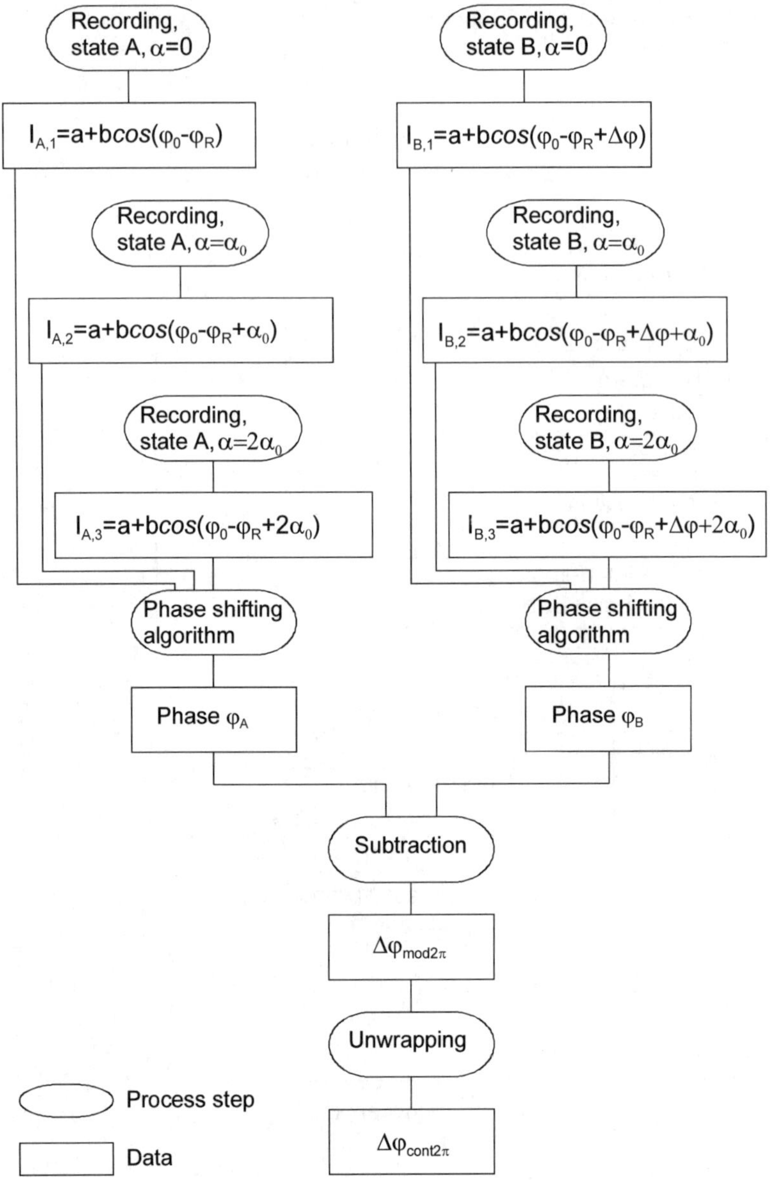


**Fig. 7.10.** Process flow of phase shifting ESPI

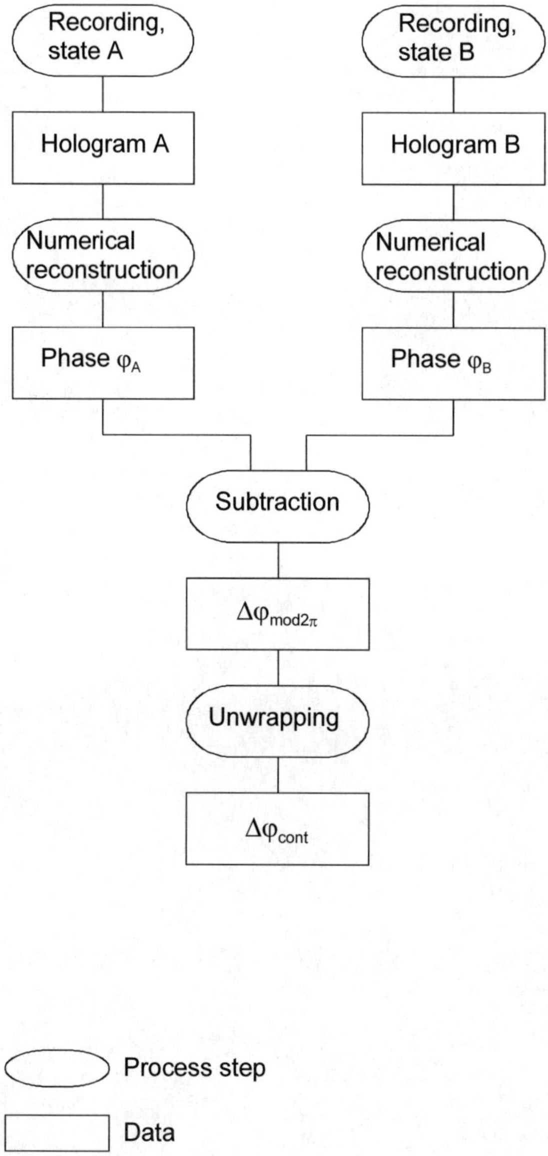


**Fig. 7.11.** Process flow of Digital HI

# Appendices

## A The Fourier Transform

### A1 Definitions

The one-dimensional *Fourier transform* of the function $f(x)$ is defined as

$$\Im\{f(x)\} = F(u) = \int_{-\infty}^{\infty} f(x)\exp[-i2\pi ux]dx \tag{A1}$$

The inverse one-dimensional *Fourier transformation* is defined as

$$\Im^{-1}\{F(u)\} = f(x) = \int_{-\infty}^{\infty} F(u)\exp[i2\pi ux]du \tag{A2}$$

The functions $f(x)$ and $F(u)$ are called *Fourier transform pair*.

The two-dimensional *Fourier transform* of the function $f(x,y)$ is defined as

$$\Im\{f(x,y)\} = F(u,v) = \int_{-\infty}^{\infty}\int_{-\infty}^{\infty} f(x,y)\exp[-i2\pi(ux+vy)]dxdy \tag{A3}$$

The corresponding inverse two-dimensional *Fourier transformation* is defined as

$$\Im^{-1}\{F(u,v)\} = f(x,y) = \int_{-\infty}^{\infty}\int_{-\infty}^{\infty} F(u,v)\exp[i2\pi(ux+vy)]dudv \tag{A4}$$

The Fourier transformation is a powerful mathematical tool to describe and analyse periodic structures. If $x$ is the time coordinate of a signal (unit: *s*), then $u$ is the corresponding frequency (unit: $1/s \equiv Hz$). In the two-dimensional case *(x,y)* are often spatial coordinates (units: *meter*), while *(u,v)* are the corresponding spatial frequencies (units: $1/meter$).

## A2 Properties

In the following some useful theorems about Fourier transforms are summarized. These formulas are written for the two-dimensional case.

1. Linearity theorem

$$\Im\{af(x,y)+bg(x,y)\}=aF(u,v)+bG(u,v) \tag{A5}$$

where $a$ and $b$ are constants, $F(u,v)=\Im(f(x,y))$ and $G(u,v)=\Im(g(x,y))$.

2. Similarity theorem

$$\Im\{f(ax,by)\}=\frac{1}{|ab|}F\left(\frac{u}{a},\frac{v}{b}\right) \tag{A6}$$

3. Shift theorem

$$\Im\{f(x-a,y-b)\}=F(u,v)\exp[-i2\pi(ua+vb)] \tag{A7}$$

4. Rayleigh's (Parseval's) theorem

$$\int_{-\infty}^{\infty}\int_{-\infty}^{\infty}|f(x,y)|^2\,dxdy=\int_{-\infty}^{\infty}\int_{-\infty}^{\infty}|F(u,v)|^2\,dudv \tag{A8}$$

5. Convolution theorem

The two-dimensional convolution of two functions $f(x,y)$ and $g(x,y)$ is defined as

$$(f\otimes g)(x,y)=\int_{-\infty}^{\infty}\int_{-\infty}^{\infty}f(x',y')g(x-x',y-y')dx'dy' \tag{A9}$$

where the $\otimes$ denotes the convolution operation. The convolution theorem states that the Fourier transform of the convolution of two functions is equal to the product of the Fourier transforms of the individual functions:

$$\Im\{f(x,y)\otimes g(x,y)\}=F(u,v)G(u,v) \tag{A10}$$

6. Autocorrelation theorem

$$\Im\left\{\int_{-\infty}^{\infty}\int_{-\infty}^{\infty}f^*(x',y')f(x+x',y+y')dx'dy'\right\}=|F(u,v)|^2 \tag{A11}$$

7. Fourier integral theorem

$$\Im\Im^{-1}\{f(x,y)\} = \Im^{-1}\Im\{f(x,y)\} = f(x,y) \tag{A12}$$

8. Differentiation
Differentiation in the spatial domain corresponds to a multiplication with a linear factor in the spatial frequency domain:

$$\Im\left\{\left(\frac{\partial}{\partial x}\right)^m \left(\frac{\partial}{\partial y}\right)^n f(x,y)\right\} = (i2\pi u)^m (i2\pi v)^n F(u,v) \tag{A13}$$

## A3 The Discrete Fourier Transform

For numerical computations the function to be transformed is given in a discrete form, i. e. $f(x)$ in Eq. (A.1) has to be replaced by the finite series $f_k$, with integer numbers $k = 0,1,...,N-1$. The continuous variable $x$ is now described as integer multiple of a sampling interval $\Delta x$ :

$$x = k\Delta x \tag{A14}$$

The frequency variable $u$ is converted into a discrete variable, too:

$$u = m\Delta u \tag{A15}$$

The discrete representation of Eq. (A.1) is then given by:

$$F_m = \Delta x \sum_{k=0}^{N-1} f_k \exp[-i2\pi km\Delta x\Delta u] \quad \text{for } m = 0,1,...,N-1 \tag{A16}$$

The maximum frequency is determined by the sampling interval in the spatial domain:

$$u_{\max} = N\Delta u = \frac{1}{\Delta x} \tag{A17}$$

The following expression

$$F_m = \frac{1}{N}\sum_{k=0}^{N-1} f_k \exp\left[-i2\pi\frac{km}{N}\right] \tag{A18}$$

is therefore defined as one-dimensional *discrete Fourier transform* (DFT). The inverse transformation is given by

$$f_k = \sum_{m=0}^{N-1} F_m \exp\left[ i2\pi \frac{km}{N} \right] \tag{A19}$$

Similar considerations lead to the discrete two-dimensional Fourier transform pair:

$$F_{mn} = \frac{1}{N^2} \sum_{k=0}^{N-1} \sum_{l=0}^{N-1} f_{kl} \exp\left[ -i2\pi \left( \frac{km + ln}{N} \right) \right] \tag{A20}$$

$$f_{kl} = \sum_{m=0}^{N-1} \sum_{n=0}^{N-1} F_{mn} \exp\left[ i2\pi \left( \frac{km + ln}{N} \right) \right] \tag{A21}$$

for $m = 0,1,...N-1$ and $n = 0,1,...N-1$

Here a quadratic field of sampling points is used, i. e. the number of points in each row is equal to that in each column.

The computation time for a discrete Fourier transform is mainly determined by the number of complex multiplications. A two-dimensional DFT can be factorised into two one-dimensional DFT's:

$$F_{mn} = \frac{1}{N^2} \sum_{k=0}^{N-1} \left[ \sum_{l=0}^{N-1} f_{kl} \exp\left( -i2\pi \frac{nl}{N} \right) \right] \exp\left( -i2\pi \frac{km}{N} \right) \tag{A22}$$

The one-dimensional DFT can be programmed most effectively using the so called *fast fourier transform* (FFT) algorithms, invented in the 70th of the last century by Cooley and Tookey. These algorithms make use of redundancies and reduce the number of multiplications for a one-dimensional DFT from $N^2$ to $2N \log_2 N$. The FFT algorithms are not described here, it is referred to the literature [10].

# B Phase Transformation of a Spherical Lens

## B1 Lens Transmission Function

The effect of an optical component with refractive index $n$ and thickness $d$ on the complex amplitude of a wave is described by a transmission function $\tau$:

$$\tau = |\tau| \exp\left[ -i\frac{2\pi}{\lambda}(n-1)d \right] \tag{B1}$$

This function is calculated in the following for a thin biconvex lens. Such lens consists of two spherical surfaces, see figure B.1. The radius of curvature of the left half lens is $R_1$, while that of the right half lens is designated $R_2$. Following sign convention is applied: As rays travel from left to right, each convex surface has a positive radius of curvature, while each concave surface has a negative radius of curvature. Due to this convention $R_2$ has a negative value. Losses due to reflection at the surfaces and due to absorption inside the lens are neglected; i. e. $|\tau| = 1$. The refractive index is constant for the entire lens.

The lens thickness is a function of the spatial coordinates $x$ and y:

$$\begin{aligned} d(x,y) &= d_1(x,y) + d_2(x,y) \\ &= d_{01} - \zeta_1 + (d_{02} - \zeta_2) \end{aligned} \tag{B2}$$

According to figure B.1 it can be written:

$$\begin{aligned} R_1^2 &= r^2 + (R_1 - \zeta_1)^2 \\ &= x^2 + y^2 + R_1^2 - 2R_1\zeta_1 + \zeta_1^2 \end{aligned} \tag{B3}$$

and

$$\begin{aligned} R_2^2 &= r^2 + (-R_2 - \zeta_2)^2 \\ &= x^2 + y^2 + R_2^2 + 2R_2\zeta_2 + \zeta_2^2 \end{aligned} \tag{B4}$$

Neglecting the quadratic terms of $\zeta_{1/2}$ leads to:

$$\zeta_1 = \frac{x^2 + y^2}{2R_1} \tag{B5}$$

$$\zeta_2 = -\frac{x^2 + y^2}{2R_2} \tag{B6}$$

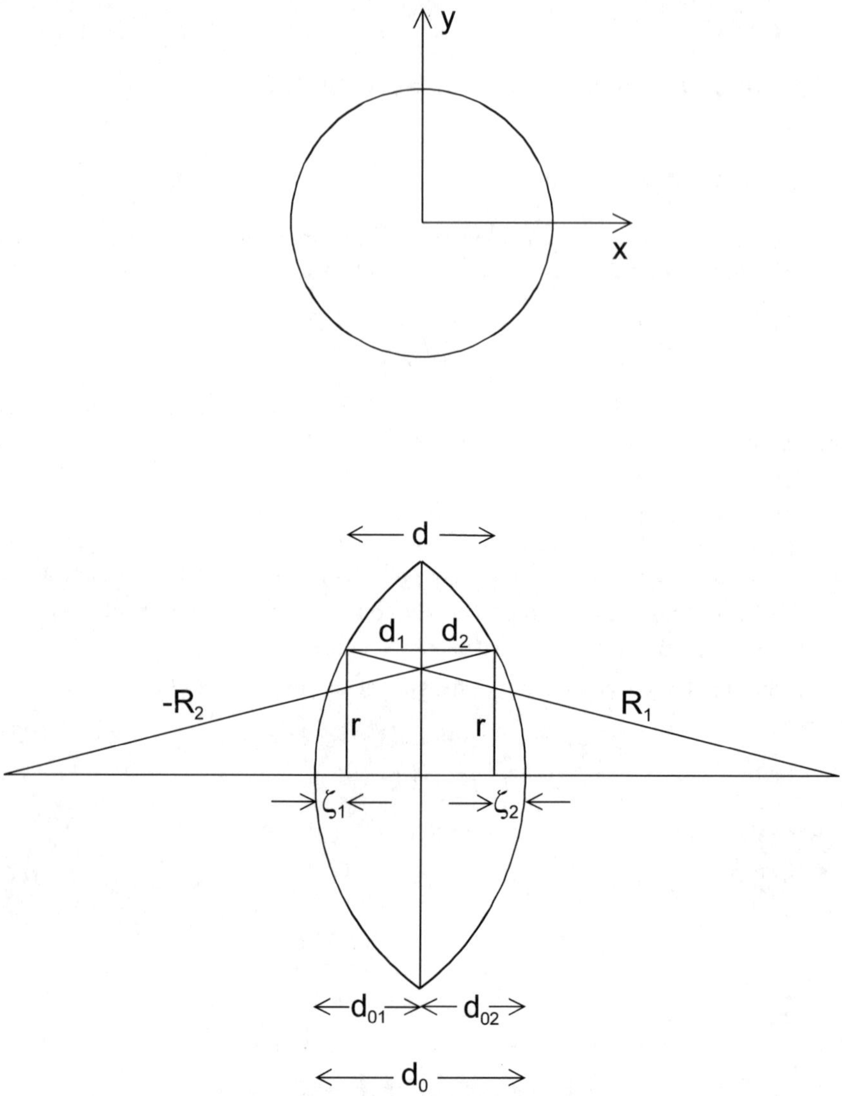


**Fig. B.1.** Biconvex lens, top view and cross-sectional view

This level of approximation is consistent with the parabolic approximation used in the Fresnel transformation. The thickness is now

$$d(x,y) = d_0 - \frac{x^2 + y^2}{2R_1} + \frac{x^2 + y^2}{2R_2} \tag{B7}$$

With the lens makers equation

$$\frac{1}{f} = (n-1)\left[\frac{1}{R_1} - \frac{1}{R_2}\right] \tag{B8}$$

of geometrical optics following lens transmission function is derived:

$$L(x,y) = \exp\left[i\frac{\pi}{\lambda f}\left(x^2 + y^2\right)\right] \tag{B9}$$

The constant factor $\exp(-i2\pi/\lambda(n-1)d_0)$, which only effects the overall phase, has been neglected.

## B2 Correction of Aberrations

In the following the complex amplitude of an object, which is imaged by a lens is calculated. The object is lying in the $(\xi,\eta)$ coordinate system, the lens is located in the (*x,y*) system and the image arises in the $(\xi',\eta')$ system, see figure B.2. The object is described by the complex amplitude $E_O(\xi,\eta)$.

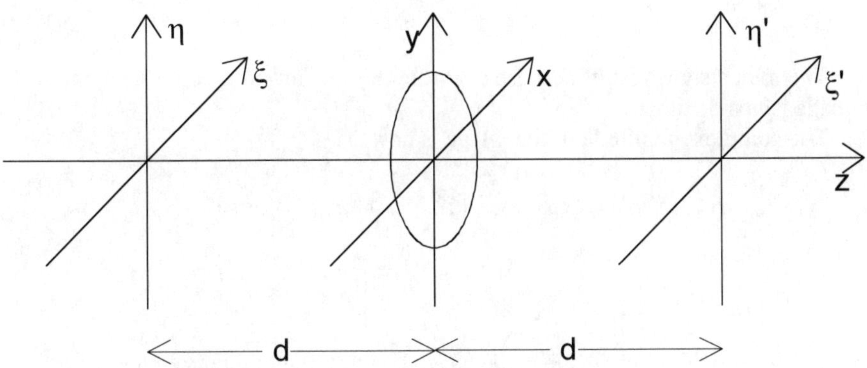


**Fig. B.2.** Image formation

The complex amplitude in front of the lens is given by

$$E_O{}'(x,y) = \exp\left[-i\frac{\pi}{\lambda d}\left(x^2 + y^2\right)\right]\int_{-\infty}^{\infty}\int_{-\infty}^{\infty} E_O(\xi,\eta)\exp\left[-i\frac{\pi}{\lambda d}\left(\xi^2 + \eta^2\right)\right] \times \exp\left[i\frac{2\pi}{\lambda d}\left(x\xi + y\eta\right)\right]d\xi d\eta \tag{B10}$$

where the Fresnel approximation is used. The complex amplitude in the image plane in then given by

$$
\begin{aligned}
&E_O''(\xi',\eta') \\
&= \exp\left[-i\frac{\pi}{\lambda d}\left(\xi'^2+\eta'^2\right)\right]\int_{-\infty}^{\infty}\int_{-\infty}^{\infty}E_O'(x,y)L(x,y)\exp\left[-i\frac{\pi}{\lambda d}\left(x^2+y^2\right)\right] \\
&\times\exp\left[i\frac{2\pi}{\lambda d}\left(x\xi'+y\eta'\right)\right]dxdy \\
&= \exp\left[-i\frac{\pi}{\lambda d}\left(\xi'^2+\eta'^2\right)\right]\int_{-\infty}^{\infty}\int_{-\infty}^{\infty}\int_{-\infty}^{\infty}\int_{-\infty}^{\infty}E_O(\xi,\eta)\exp\left[-i\frac{\pi}{\lambda d}\left(x^2+y^2\right)\right] \\
&\times\exp\left[-i\frac{\pi}{\lambda d}\left(\xi^2+\eta^2\right)\right]\exp\left[i\frac{2\pi}{\lambda d}\left(x\xi+y\eta\right)\right]\exp\left[i\frac{2\pi}{\lambda d}\left(x^2+y^2\right)\right] \\
&\times\exp\left[-i\frac{\pi}{\lambda d}\left(x^2+y^2\right)\right]\exp\left[i\frac{2\pi}{\lambda d}\left(x\xi'+y\eta'\right)\right]d\xi d\eta dxdy
\end{aligned}
\tag{B11}
$$

A magnification of 1 and a focal distance of $f = d/2$ is used for the lens transmission function $L(x,y)$.

The image coordinates can be expressed in terms of the object coordinates:

$$\xi' = -\xi \quad \text{and} \quad \eta' = -\eta \tag{B12}$$

The minus signs result, because according to the laws of geometrical optics the image is upside down.

The complex amplitude of the image is now

$$
\begin{aligned}
E_O''(\xi',\eta') &= \exp\left[-i\frac{2\pi}{\lambda d}\left(\xi'^2+\eta'^2\right)\right]E_O(-\xi',-\eta') \\
&= \exp\left[-i\frac{\pi}{\lambda f}\left(\xi'^2+\eta'^2\right)\right]E_O(-\xi',-\eta')
\end{aligned}
\tag{B13}
$$

The wavefield in the image plane has to be multiplied therefore by a factor

$$P(\xi',\eta') = \exp\left[i\frac{\pi}{\lambda f}\left(\xi'^2+\eta'^2\right)\right] \tag{B14}$$

in order to generate the correct phase distribution.

This correction factor depends on the wavelength and on the coordinates of the image plane. It can be neglected, if only the intensity of a wavefield has to be calculated after reconstruction ($I \propto P^*P = 1$). This is also valid, if the phase difference of two wavefields, which are recorded with the *same* wavelength, is computed:

$$
\begin{aligned}
&\Delta\varphi = \varphi_1 - \varphi_2 = i\pi/\lambda f\left(\xi'^2+\eta'^2\right)+\varphi_1' - \left[i\pi/\lambda f\left(\xi'^2+\eta'^2\right)+\varphi_2'\right] \\
&= \varphi_1' - \varphi_2'
\end{aligned}
\tag{B15}
$$

This is usually the case in DHI for applications in deformation analysis. However, the correction factor has to be considered, if the phase difference of two wavefields, which are recorded with *different* wavelengths, is computed. This is the case in multi-wavelength DHI for shape measurement.

# References


1. Abramson N (1983) Light-in-flight recording: High speed holographic motion pictures of ultrafast phenomena. Appl Opt 22:215-232
2. Abramson N (1984) Light-in-flight recording 2: Compensation for the limited speed of the light used for observation. Appl Opt 23:1481-1492
3. Abramson N (1984) Light-in-flight recording 3: Compensation for optical relativistic effects. Appl Opt 23:4007-4014
4. Abramson N (1985) Light-in-flight recording 4: Visualizing optical relativistic phenomena. Appl Opt 24:3323-3329
5. Adams M, Kreis T, Jüptner W (1997) Particle size and position measurement with digital holography. In: Proc SPIE vol 3098, pp 234-240
6. Adams M, Kreis T, Jüptner W (1999) Particle measurement with digital holography. In: Proc SPIE vol 3823
7. Bartels RA, Paul A, Green H, Kapteyn HC, Murnane MM, Backus S, Christov IP, Liu Y, Attwood D, Jacobsen C (2002) Generation of Spatially Coherent Light at Extreme Ultraviolet Wavelengths. Science 297:376-378
8. Bisle W (1998) Optische Prüfung an Luftfahrtkomponenten: Weiterentwicklung des Scherografie-Prüfverfahrens für nicht-kooperative Oberflächen von Flugzeugstrukturen. Proc Deutsche Gesellschaft für Luft- und Raumfahrt, annual meeting, Bremen
9. Breuckmann B, Thieme W (1985) Computer-aided analysis of holographic interferograms using the phase-shift method. Appl Opt 24:2145-2149
10. Brigham EO (1974) The fast fourier transform. Pretince-Hall
11. Bryngdahl O, Wyrowski F (1990) Digital Holography - computer generated holograms. Progress in Optics 28:1-86
12. Butters JN, Leendertz JA (1971) Holographic and Videotechniques applied to engineering measurements. J Meas Control 4:349-354
13. Carlsson T, Nilsson B, Gustafsson J (2001) System for acquisition of three-dimensional shape and movement using digital Light-in-Flight holography. Opt Eng 40(01): 67-75
14. CCD Primer (2002) product information. Kodak
15. Chen DJ, Chiang FP, Tan YS, Don HS (1993) Digital speckle displacement measurement using a complex spectrum method. Appl Opt 32(11): 1839-1848
16. Coppola G, De Nicola S, Ferraro P, Finizio A, Grilli S, Iodice M, Magro C, Pierattini G (2003) Evaluation of residual stress in MEMS structures by means of digital holography. In: Proc. SPIE vol 4933, pp 226-31
17. Coppola G, Ferraro P, Iodice M, De Nicola S, Finizio A, Grilli S (2004) A digital holographic microscope for complete characterization of microelectromechanical systems. Meas Sci Technol 15: 529-539
18. Coquoz O, Conde R, Taleblou F, Depeursinge C (1995) Performances of endoscopic holography with a multicore optical fiber. Appl Opt 34(31):7186-7193
19. Creath K (1985) Phase shifting speckle-interferometry. Appl Opt 24(18):3053-3058

20. Creath K (1994) Phase-shifting holographic interferometry. In: Holographic Interferometry, Springer Series in Optical Sciences 68, pp 109-150
21. Cuche E, Bevilacqua F, Depeursinge C (1999) Digital holography for quantitative phase-contrast imaging. Optics Letters 24(5):291-293
22. Cuche E, Marquet P, Depeursinge C (1999) Simultaneous amplitude-contrast and quantitative phase-contrast microscopy by numerical reconstruction of Fresnel off-axis holograms. Appl Opt 38(34):6994-7001
23. Cuche E, Marquet P, Depeursinge C (2000) Spatial filtering for zero-order and twin-image elimination in digital off-axis holography. Appl Opt 39(23):4070-4075
24. Cuche E, Marquet P, Depeursinge C (2000) Aperture apodization using cubic spline interpolation: application in digital holographic microscopy. Opt Commun 182:59-69
25. Demetrakopoulos TH, Mittra R (1974) Digital and optical reconstruction of images from suboptical diffraction patterns. Appl Opt 13(3):665-670
26. Demoli N, Mestrovic J, Sovic I (2003) Subtraction digital holography. Appl Opt 42(5):798-804
27. De Nicola S, Ferraro P, Finizio A, Pierattini G (2001) Correct-image reconstruction in the presence of severe anamorphism by means of digital holography. Opt Letters 26(13):974-76
28. De Nicola S, Ferraro P, Finizio A, Pierattini G (2001) Compensation of Aberrations in Fresnel off-axis Digital Holography. In: Jüptner W, Osten W (eds) Proc 4th International Workshop on Automatic Processing of Fringe Patterns. Akademie, Berlin, pp 407-412
29. Doval AF (2000) A systematic approach to TV holography. Meas Sci Technol 11:R1-R36
30. Dubois F, Joannes L, Legros J C (1999) Improved three-dimensional imaging with a digital holography microscope with a source of partial spatial coherence. Appl Opt 38(34):7085-7094
31. Dubois F, Joannes L, Dupont O, Dewandel JL, Logros JC (1999) An integrated optical set-up for fluid-physics experiments under microgravity conditions. Meas Sci Technol 10:934-945
32. Dubois F, Minetti C, Monnom O, Yourassowsky C, Legros JC, Kischel P (2002) Pattern recognition with a digital holographic microscope working in partially coherent illumination. Appl Opt 41(20):4108-4119
33. Dubois F, Monnom O, Yourassowsky C, Legros JC (2002) Border processing in digital holography by extension of the digital hologram and reduction of the higher spatial frequencies. Appl Opt 41(14):2621-2626
34. Frauel Y, Javidi B (2001) Neural network for three-dimensional object recognition based on digital holography. Optics Letters 26(19):1478-1480
35. Frauel Y, Tajahuerce E, Castro MA, Javidi B (2001) Distortion- tolerant three-dimensional object recognition with digital holography. Appl Opt 40(23):3887
36. Füzessy Z, Gyimesi F (1984) Difference holographic interferometry: displacement measurement. Opt Eng 23(6):780-783
37. Gabor D (1948) A new microscopic principle. Nature 161:777-778
38. Gabor D (1949) Microscopy by reconstructed wavefronts. Proc Roy Soc 197:454-487
39. Gabor D (1951) Microscopy by reconstructed wavefronts: 2. Proc Phys Soc 64:449-469
40. Goodman JW (1975) statistical properties of laser speckle patters. In: Dainty JC (ed) Laser Speckle and Related Phenomena, Topics in Appl Physics vol 9, Springer, Berlin, pp. 9-75
41. Goodman JW (1996) Introduction to Fourier Optics 2nd ed. McGraw-Hill

42. Goodman JW, Lawrence RW (1967) Digital image formation from electronically detected holograms. Appl Phys Lett 11:77-79
43. Grilli S, Ferraro P, De Nicola S, Finizio A, Pierattini G Meucci R (2001) Whole optical wavefield reconstruction by digital holography. Optics Express 9(6):294-302
44. Grunwald R, Griebner U, Elsaesser T, Kebbel V, Hartmann H J, Jüptner W (2001) Femtosecond interference experiments with thin-film micro-optical components. In: Jüptner W, Osten W (eds) Proc 4th International Workshop on Automatic Processing of Fringe Patterns. Akademie, Berlin, pp 33-40
45. Guo CS, Zhang L, Wang HT, Liao J, Zhu Y (2002) Phase-shifting error and its elimination in phase-shifting digital holography. Opt Lett 27(19):1687-1689
46. Haddad W, Cullen D, Solem J, Longworth J, McPherson A, Boyer K, Rhodes K (1992) Fourier-transform holographic microscope. Appl Opt 31(24): 4973-4978
47. Harriharan P (1984) Optical Holography. Cambridge University Press, Cambridge
48. Helmers H, Bischoff M, Ehlkes L (2001)ESPI-System with active in-line digital phase stabilization. In: Jüptner W, Osten W (eds) Proc 4th International Workshop on Automatic Processing of Fringe Patterns. Elsevier, pp 673-679
49. Hinsch K (2002) Holographic particle image velocimetry. Meas Sci Technol 13:R61-R72
50. Holstein D, Hartmann HJ, Jüptner W (1998) Investigation of Laser Welds by Means of Digital Speckle Photography. In: Proc SPIE vol 3478, pp 294-301
51. Hung YY (1996) Shearography for non-destructive evaluation of composite structures. Opt Lasers Eng 24:161-182
52. Hung YY, Liang CY (1979) Image shearing camera for direct measurement of surface strain. Appl Opt 18:1046-1051
53. Huntley JM, Saldner H (1993) Temporal phase-unwrapping algorithm for automated interferogram analysis. Appl Opt 32(17):3047-3052
54. Inomato O, Yamaguchi I 2001 Measurements of Benard-Marangoni waves using phase-shifting digital holography. In: Proc SPIE vol 4416:124-127
55. Jacquot M, Sandoz P, Tribillon G (2001) High resolution digital holography. Opt Commun 190:87-94
56. Javidi B, Nomura T (2000) Securing information by use of digital holography. Optics Letters 25(1):28-30
57. Javidi B, Tajahuerce E (2000) Three-dimensional object recognition by use of digital holography. Optics Letters 25(9):610-612
58. Jüptner W (1978) Automatisierte Auswertung holografischer Interferogramme mit dem Zeilen-Scanverfahren. In: Kreitlow H, Jüptner W (eds) Proc Frühjahrsschule 78 Holografische Interferometrie in Technik und Medizin.
59. Jüptner W (2000) Qualität durch Lasertechnik - Zukunft für das 21. Jahrhundert. In: Proc LEF symposium. Erlangen
60. Jüptner W, Kreis T, Kreitlow H (1983) Automatic evaluation of holographic interferograms by reference beam phase shifting. In: Proc SPIE vol 398, pp 22-29
61. Jüptner W, Pomarico J, Schnars U (1996) Light-in-Flight measurements by Digital Holography. In: Proc. SPIE vol 2860
62. Kato J, Yamaguchi I, Matsumura T (2002) Multicolor digital holography with an achromatic phase shifter. Opt Lett 27(16):1403-1405
63. Kebbel V, Grubert B, Hartmann HJ, Jüptner W, Schnars U (1998) Application of Digital Holography to Space-Borne Fluid Science Measurements. In: Proc 49$^{th}$ International Astronautical Congress Melbourne paper no. IAF-98-J.5.03
64. Kebbel V, Adams M, Hartmann H J, Jüptner W (1999) Digital holography as a versatile optical diagnostic method for microgravity experiments. Meas Sci Technol 10:893-899

65. Kebbel V, Hartmann HJ, Jüptner W (2001) Application of digital holographic microscopy for inspection of micro-optical components. In: Proc SPIE vol 4398, pp 189-98
66. Kim MK (1999) Wavelength-scanning digital interference holography for optical section imaging. Optics Letters 24(23):1693-1695
67. Kim MK (2000) Tomographic three-dimensional imaging of a biological specimen using wavelength-scanning digital interference holography. (2000) Optics Express 7(9):305-310
68. Kim S, Lee B, Kim E (1997) Removal of bias and the conjugate image in incoherent on-axis triangular holography and real-time reconstruction of the complex hologram. Appl Opt 36(20):4784-4791
69. Klein MV, Furtak TE (1986) Optics. 2nd ed Wiley, New York
70. Kolenovic E, Lai S, Osten W, Jüptner W (2001) Endoscopic shape and deformation measurement by means of Digital Holography. In: Jüptner W, Osten W (eds) Proc 4th International Workshop on Automatic Processing of Fringe Patterns. Akademie, Berlin, pp 686-691
71. Kreis T (1996) Holographic Interferometry. Akademie, Berlin
72. Kreis T (2002) Frequency analysis of digital holography. Opt Eng 41(4):771-778
73. Kreis T (2002) Frequency analysis of digital holography with reconstruction by convolution. Opt Eng 41(8):1829-1839
74. Kreis T, Jüptner W (1997) Principles of digital holography. In: Jüptner W, Osten W (eds) Proc 3rd International Workshop on Automatic Processing of Fringe Patterns. Akademie, Berlin, pp 353-363
75. Kreis T, Jüptner W (1997) Suppression of the dc term in digital holography. Opt Eng 36(8):2357-2360
76. Kreis T, Jüptner W, Geldmacher J (1998) Digital Holography: Methods and Applications. In: Proc SPIE vol 3407, pp 169-177
77. Kreis T, Adams M, Jüptner W 1999 Digital in-line holography in particle measurement. In: Proc SPIE vol 3744
78. Kreis T, Aswendt P, Höfling R (2001) Hologram reconstruction using a digital micromirror device. Opt Eng 40(6):926-933
79. Kreis T, Adams M, Jüptner W (2002) Aperture synthesis in digital holography. In: Proc SPIE vol 4777, pp 69-76
80. Kreuzer HJ, Pawlitzek RA (1997) Numerical Reconstruction for in-line Holography in Reflection and under glancing Incidence. In: Jüptner W, Osten W (eds) Proc 3rd International Workshop on Automatic Processing of Fringe Patterns. Akademie, Berlin, pp 364-367
81. Kronrod MA, Merzlyakov NS, Yaroslavski LP (1972) Reconstruction of holograms with a computer. Sov Phys-Tech Phys USA 17 (2):333-334
82. Krupka R, Walz T, Ettemeyer A (2002) Industrial Applications of Shearography for Inspection of Aircraft Components. Proc 8$^{th}$ ECNDT, Barcelona
83. Kulak M, Pisarek J (2001) Speckle photography in the examination of composites. In: Jüptner W, Osten W (eds) Proc 4th International Workshop on Automatic Processing of Fringe Patterns. Elsevier, pp 528-530
84. Lai S, Neifeld M (2000) Digital wavefront reconstruction and its application to image encryption. Opt Commun 178:283-289
85. Lai S, Kemper B, von Bally G (1999) Off-axis reconstruction of in-line holograms for twin-image elimination. Optics Communications 169:37-43
86. Lai S, King B, Neifeld M A (2000) Wave front reconstruction by means of phase-shifting digital in-line holography. Optics Communications 173:155-160

87. Lai S, Kolenovic E, Osten W, Jüptner W (2002) A deformation and 3D-shape measurement system based on phase-shifting digital holography. In: Proc SPIE vol 4537, pp 273-276
88. Le Clerc F, Gross M (2001) Synthetic-aperture experiment in the visible with on-axis digital heterodyne holography. Opt Lett 26(20):1550-1552
89. Le Clerc F, Collot L, Gross M (2000) Numerical heterodyne holography with two-dimensional photodetector arrays. Optics Letters 25(10):716-718
90. Lee WH (1978) Computer-generated Holograms: Techniques and Applications. Progress in Optics 16: 120-232
91. Leith EN, Upatnieks J (1962) Reconstructed wavefronts and communication theory. Jour Opt Soc Amer 52:1123-1130
92. Leith EN, Upatnieks J (1964) Wavefront reconstruction with diffused illumination and threedimensional objects. Journ Opt Soc Amer 54:1295-1301
93. Liu G, Scott PD (1987) Phase retrieval and twin-image elimination for in-line Fresnel holograms. J Opt Soc Am A 4(1): 159-165
94. Lokberg O (1980) Electronic Speckle Pattern Interferometry. Phys Techn 11:16-22
95. Lokberg O, Slettemoen GA (1987) Basic Electronic Speckle Pattern Interferometry. Applied Optics and Optical Engineering 10:455-505
96. Macovski A, Ramsey D, Schaefer LF (1971) Time Lapse Interferometry and Contouring using Television Systems. Appl. Opt. 10(12):2722-2727
97. Matoba O, Naughton TJ, Frauel Y, Bertaux N, Javidi B (2002) Real-time three-dimensional object reconstruction by use of a phase-encoded digital hologram. Appl Opt 41(29):6187-6192
98. Milgram JH, Weichang Li (2002) Computational reconstruction of images from holograms. Appl Opt 41(5): 853-864
99. Nadeborn W, Andrä P, Osten W (1995) A robust procedure for absolute phase measurement. Optics and Lasers in Engineering 22
100. Neumann DB (1980) Comparative Holography. In: Tech Digest Topical Meeting on Hologram Interferometry and Speckle Metrology, paper MB2-1. Opt Soc Am
101. Nilsson B, Carlsson T (1998) Direct three-dimensional shape measurement by digital light-in-flight holography. Appl Opt 37(34):7954-7959
102. Nilsson B, Carlsson T (1999) Digital light-in-flight holography for simultaneous shape and deformation measurement. In: Proc. SPIE vol 3835, pp 127-134
103. Nilsson B, Carlsson T 2000 Simultaneous measurement of shape and deformation using digital light-in-flight recording by holography. Opt Eng 39(1):244-253
104. Onural L (2000) Sampling of the diffraction field. Appl Opt 39(32):5929-5935
105. Onural L, Özgen MT (1992) Extraction of three-dimensional object-location information directly from in-line holograms using Wigner analysis. J Opt Soc Amer A 9(2):252-260
106. Onural L, Scott PD (1987) Digital decoding of in-line holograms. Opt Eng 26(11):1124-1132
107. Osten W, Nadeborn W, Andrä P (1996) General hierarchical approach in absolute phase measurement. In: Proc SPIE vol 2860
108. Osten W, Kalms M, Jüptner, Tober G, Bisle W, Scherling D (2000) Shearography system for the testing of large scale aircraft components taking into account noncooperative surfaces. In: Proc SPIE vol 4101B
109. Osten W, Seebacher S, Jüptner W (2001) Application of digital holography for the inspection of microcomponents. In: Proc SPIE vol 4400, pp 1-15
110. Osten W, Seebacher S, Baumbach T, Jüptner W (2001) Absolute shape control of microcomponents using digital holography and multiwavelength contouring. In: Proc SPIE vol 4275, pp 71-84

111. Osten W, Baumbach T, Seebacher S, Jüptner W (2001) Remote shape control by comparative digital holography. In: Jüptner W, Osten W (eds) Proc 4th International Workshop on Automatic Processing of Fringe Patterns. Akademie, Berlin, pp 373-382
112. Osten W, Baumbach T, Jüptner W (2002) Comparative digital holography. Opt Lett 27(20):1764-1766
113. Ostrovsky YI, Butosov MM, Ostrovskaja GV (1980) Interferometry by Holography. Springer, New York
114. Owen R B, Zozulya A (2000) In-line digital holographic sensor for monitoring and characterizing marine particulates. Opt Eng 39(8):2187-2197
115. Owen RB, Zozulya A, Benoit MR, Klaus DM (2002) Microgravity materials and life sciences research applications of digital holography. Appl Opt 41(19): 3927-3935
116. Papp Z, Janos K (2001) Digital holography by two reference beams. In: Proc SPIE vol 4416, pp 112-115
117. Pedrini G, Tiziani H (2002) Short-coherence digital microscopy by use of a holographic imaging system. Appl Opt 41(22):4489-4496
118. Pedrini G, Zou YL, Tiziani H (1995) Digital double-pulsed holographic interferometry for vibration analysis. J Mod Opt 42(2):367-374
119. Pedrini G, Zou Y, Tiziani H (1997) Simultaneous quantitative evaluation of in-plane and out-of-plane deformations by use of a multidirectional spatial carrier. Appl Opt 36(4):786
120. Pedrini G, Fröning P, Fessler H, Tiziani HJ (1998) In-line digital holographic interferometry. Appl Opt 37(26):6262-6269
121. Pedrini G, Schedin S, Tiziani H (1999) Lensless digital holographic interferometry for the measurement of large objects. Optics Communications 171:29-36
122. Pedrini G, Fröning P, Tiziani H, Santoyo F (1999) Shape measurement of microscopic structures using digital holograms. Opt Commun 164:257-268
123. Pedrini G, Schedin S, Tiziani H (2000) Spatial filtering in digital holographic microscopy. J Mod Opt 47(8):1447-1454
124. Pedrini G, Titiani H J, Alexeenko I (2002) Digital-holographic interferometry with an image-intensifier system. Appl Opt 41(4):648
125. Pettersson S-G, Bergstrom H, Abramson N (1989) Light-in-flight recording 6: Experiment with view-time expansion using a skew reference wave. Appl Opt 28:766-770
126. Pomarico J, Schnars U, Hartmann HJ, Jüptner W (1996) Digital recording and numerical reconstruction of holograms: A new method for displaying Light-in-flight. Applied Optics 34(35):8095-8099
127. Powell RL, Stetson KA (1965) Interferometric Vibration Analysis by Wavefront reconstructions. J Opt Soc Amer 55:1593-1598
128. Schnars U (1994) Direct phase determination in hologram interferometry with use of digitally recorded holograms. Journ Opt Soc Am A 11(7):2011-2015, reprinted (1997) In: Hinsch K, Sirohi R (eds). SPIE Milestone Series MS 144, pp 661 - 665
129. Schnars U (1994) Digitale Aufzeichnung und mathematische Rekonstruktion von Hologrammen in der Interferometrie VDI-Fortschritt-Berichte series 8 no 378 VDI, Düsseldorf
130. Schnars U, Jüptner W (1993) Principles of direct holography for interferometry. In: Jüptner W, Osten W (eds) FRINGE 93 Proc. 2nd International Workshop on Automatic Processing of Fringe Patterns. Akademie, Berlin, pp 115-120
131. Schnars U, Jüptner W (1994) Direct recording of holograms by a CCD-target and numerical reconstruction. Applied Optics 33(2):179-181

132. Schnars U, Jüptner W (1994) Digital reconstruction of holograms in hologram interferometry and shearography. Applied Optics 33(20):4373-4377, reprinted (1997) In: Hinsch K, Sirohi R (eds). SPIE Milestone Series MS 144, pp 656 - 660
133. Schnars U, Jüptner W (1995) Digitale Holografie. In: annual conference of the Deutsche Gesellschaft für angewandte Optik. Handout, Binz
134. Schnars U, Geldmacher J, Hartmann HJ, Jüptner W (1995) Mit digitaler Holografie den Stoßwellen auf der Spur. F&M 103(6):338-341
135. Schnars U, Kreis T, Jüptner W (1995) CCD recording and numerical reconstruction of holograms and holographic interferograms. In: Proc SPIE vol 2544, pp 57-63
136. Schnars U, Kreis T, Jüptner W (1995) Numerische Rekonstruktion von Hologrammen in der interferometrischen Messtechnik, In: Waidelich W (ed) Proc LASER 95. Springer, Heidelberg
137. Schnars U, Osten W, Jüptner W, Sommer K (1995) Advances of Digital Holography for Experiment Diagnostics in Space. In: Proc 46$^{th}$ International Astronautical Congress Oslo, paper no. IAF-95-J.5.01
138. Schnars U, Hartmann HJ, Jüptner W (1995) Digital recording and numerical reconstruction of holograms for nondestructive testing. In: Proc SPIE vol 2545, pp 250-253
139. Schnars U, Kreis T, Jüptner W (1996) Digital recording and numerical reconstruction of holograms: Reduction of the spatial frequency spectrum. Optical Engineering 35(4):977-982
140. Schreier D (1984) Synthetische Holografie. VCH, Weinheim
141. Schwomma O (1972) austrian patent 298,830
142. Seebacher S (2001) Anwendung der digitalen Holografie bei der 3D-Form- und Verformungsmessung an Komponenten der Mikrosystemtechnik. University Bremen publishing house, Bremen
143. Seebacher S, Osten W, Jüptner W (1998) Measuring shape and deformation of small objects using digital holography. In: Proc SPIE vol 3479, pp 104-115
144. Seebacher S, Baumbach T, Osten W, Jüptner W (2000) Combined 3D-shape and deformation analysis of small objects using coherent optical techniques on the basis of digital holography. In: Proc SPIE vol 4101B, pp 520-531
145. Sjoedahl M, Benckert L R (1993) Electronic speckle photography: analysis of an algorithm giving the displacement with subpixel accuracy. Appl Opt 32(13):2278-2284
146. Skarman B, Becker J, Wozniak K (1996) Simultaneous 3D-PIV and temperature measurements using a new CCD-based holographic interferometer. Flow Meas Instrum 7(1):1-6
147. Sollid J E (1969) Holographic interferometry applied to measurements of small static displacements of diffusely reflecting surfaces. Appl Opt 8:1587-1595
148. Stadelmaier A, Massig JH (2000) Compensation of lens aberrations in digital holography. Optics Letters 25(22):1630-1632
149. Steinbichler H (2004) Shearography – NDT. Product information, Steinbichler, Neubeuern
150. Steinchen W, Yang L (2003) Digital Shearography. SPIE press
151. Stetson KA, Powell RL (1965) Interferometric hologram evaluation and real-time vibration analysis of diffuse objects. J Opt Soc Amer 55:1694-1695
152. Stetson KA, Brohinsky R (1985) Electrooptic holography and its application to hologram interferometry. Appl Opt 24(21):3631-3637
153. Stetson KA, Brohinsky R (1987) Electrooptic holography system for vibration analysis and nondestructive testing. Opt Eng 26(12):1234-1239
154. Synnergren P, Sjödahl M (2000) Mechanical testing using digital speckle photography. Proc SPIE vol 4101B

155. Tajahuerce E, Javidi B (2000) Encrypting three-dimensional information with digital holography. Appl Opt 39(35):6595-6601
156. Tajahuerce E, Matoba O, Verral S, Javidi B (2000) Optoelectronic information encryption with phase-shifting interferometry. Appl Opt 39(14):2313-2320
157. Tajahuerce E, Matoba O, Javidi B (2001) Shift-invariant three-dimensional object recognition by means of digital holography. Appl Opt 40(23):3877-3886
158. Takaki Y, Ohzu H (1999) Fast numerical reconstruction technique for high-resolution hybrid holographic microscopy. Appl Opt 38(11):2204-2211
159. Takaki Y, Ohzu H (2000) Hybrid holographic microscopy: visualization of three-dimensional object information by use of viewing angles. Appl Opt 39(29):5302-5308
160. Takaki Y, Kawai H, Ohzu H (1999) Hybrid holographic microscopy free of conjugate and zero-order images. Appl Opt 38(23):4990-4996
161. Trolinger JD (1991) Particle and Flow Field Holography Combustion Measurements. Chigier N (ed). Hemisphere Publishing Corporation, pp 51-89
162. Wagner C, Seebacher S, Osten W, Jüptner W (1999) Digital recording and numerical reconstruction of lensless Fourier holograms in optical metrology. Appl Opt 38(22):4812-4820
163. Wagner C, Osten W, Seebacher S (2000) Direct shape measurement by digital wavefront reconstruction and multiwavelength contouring. Opt Eng 39(1):79-85
164. Winnacker A (1984) Physik von Laser und Maser. BI-Verlag, Mannheim
165. Wozniak K, Skarman B (1994) Digital Holography in Flow Visualization. Final Report for ESA/ ESTEC purchase order 142722, Noordwijk
166. Xiao X, Puri I (2002) Digital recording and numerical reconstruction of holograms: an optical diagnostic for combustion. Appl Opt 41(19):3890-3898
167. Xu L, Miao J, Asundi A (2000) Properties of digital holography based on in-line configuration. Opt Eng 39(12):3214-3219
168. Xu L, Peng X, Miao J, Asundi A K (2001) Studies of digital microscopic holography with applications to microstructure testing. Appl Opt 40(28):5046-5051
169. Xu L, Peng X, Asundi A, Miao J, (2001) Hybrid holographic microscope for interferometric measurement of microstructures. Opt Eng 40(11):2533-2539
170. Yamaguchi I, Saito H (1969) Application of holographic interferometry to the measurement of poisson´s ratio. Jap Journal of Appl Phys 8:768-771
171. Yamaguchi I , Zhang T (1997) Phase-shifting digital holography. Optics Letters 22(16):1268-1270
172. Yamaguchi I, Kato J, Ohta S, Mizuno J (2001) Image formation in phase-shifting digital holography and applications to microscopy. Appl Opt 40(34):6177-6186
173. Yamaguchi I, Inomoto O, Kato J (2001) Surface shape measurement by phase shifting digital holography. In: Jüptner W, Osten W (eds) Proc 4th International Workshop on Automatic Processing of Fringe Patterns. Akademie, Berlin, pp 365-372
174. Yamaguchi I, Matsumura T, Kato J (2002) Phase-shifting color digital holography. Opt Lett 27(13):1108-1110
175. Yang S, Xie X, Thuo Y, Jia C (1999) Reconstruction of near-field in-line holograms. Optics Communications 159:29-31
176. Yaroslavskii LP, Merzlyakov NS (1980) Methods of digital holography. Consultants Bureau, New York
177. Yu L, Cai L (2001) Iterative algorithm with a constraint condition for numerical reconstruction of a three-dimensional object from its hologram. J Opt Soc Am A 18(5):1033-1045
178. Zhang T, Yamaguchi I (1998) Three-dimensional microscopy with phase-shifting digital holography. Optics Letters 23(15):1221-1223

179. Zhang T, Yamaguchi I (1998) 3D microscopy with phase-shifting digital holography. In: Proc SPIE vol 3479:152-159
180. Zou Y, Pedrini G, Tiziani H (1996) Surface contouring in a video frame by changing the wavelength of a diode laser. Opt Eng 35(4):1074-1079

# Index

Printing: Krips bv, Meppel
Binding: Litges & Dopf, Heppenheim